SpringerBriefs in Molecular Science

SpringerBriefs in Molecular Science present concise summaries of cutting-edge research and practical applications across a wide spectrum of fields centered around chemistry. Featuring compact volumes of 50 to 125 pages, the series covers a range of content from professional to academic. Typical topics might include:

- A timely report of state-of-the-art analytical techniques
- A bridge between new research results, as published in journal articles, and a contextual literature review
- A snapshot of a hot or emerging topic
- An in-depth case study
- A presentation of core concepts that students must understand in order to make independent contributions

Briefs allow authors to present their ideas and readers to absorb them with minimal time investment. Briefs will be published as part of Springer's eBook collection, with millions of users worldwide. In addition, Briefs will be available for individual print and electronic purchase. Briefs are characterized by fast, global electronic dissemination, standard publishing contracts, easy-to-use manuscript preparation and formatting guidelines, and expedited production schedules. Both solicited and unsolicited manuscripts are considered for publication in this series.

More information about this series at http://www.springer.com/series/8898

Masahiro Kinoshita

Mechanism of Functional Expression of F_1-ATPase

 Springer

Masahiro Kinoshita
Institute of Advanced Energy
Kyoto University
Uji, Kyoto, Japan

Graduate School of Science
Chiba University
Chiba, Japan

ISSN 2191-5407   ISSN 2191-5415   (electronic)
SpringerBriefs in Molecular Science
ISBN 978-981-33-6234-5   ISBN 978-981-33-6232-1   (eBook)
https://doi.org/10.1007/978-981-33-6232-1

This Springer imprint is published by the registered company Springer Nature Singapore Pte Ltd.
The registered company address is: 152 Beach Road, #21-01/04 Gateway East, Singapore 189721, Singapore

Preface

In this book, we present a view on the mechanism of functional expression of an ATP-driven molecular motor (a protein or protein complex). It is substantially different from the prevailing view that the molecular motor converts the chemical energy stored in an ATP molecule or the free energy of ATP hydrolysis reaction to mechanical work. We point out that this concept of chemo-mechanical coupling is problematic and inconsistent with some of the recent experimental observations. The molecular motor acts as a catalyst for the ATP hydrolysis reaction, an irreversible process, with the result that it is involved in the ATP hydrolysis cycle comprising the ATP binding to the molecular motor, ATP hydrolysis, and dissociation of ADP and Pi from the molecular motor. During this cycle, the molecular motor exhibits sequential changes in its configuration for minimizing the system free energy. Here, the system comprises not only the molecular motor but also water in which ATP, ADP, and Pi are dissolved. The system performs essentially no mechanical work. The force which moves the protein or rotates a protein in the complex is generated by not ATP but water. The effect of translational displacement of water molecules in the entire system makes a dominant contribution to the force generation. On the basis of the new view in which this entropic force by water plays a pivotal role, we unveil the mechanism of unidirectional rotation of the central shaft in F_1-ATPase. This book describes the latest ideas of the author which have not been published yet.

Uji/Chiba, Japan   Masahiro Kinoshita

Acknowledgments

The author thanks all the collaborators listed in the References, in particular, Takashi Yoshidome and Mitsunori Ikeguchi, his major collaborators for F_1-ATPase. Appreciation should be expressed to Takeshi Murata, Tomohiko Hayashi, and Satoshi Yasuda for the critical reading of the original draft. The author wishes to express his deep gratitude to Eiro Muneyuki who kindly answered a number of questions on F_1-ATPase from him. Useful comments from Makoto Suzuki, Akira Yoshimori, and Mitsuhiro Iwaki are truly acknowledged. Appreciation is also due to Satoshi Yasuda who prepared Figs. 3.3 and 3.12. The studies described in this book were supported by Grant-in-Aid for Scientific Research on Innovative Areas (No. 20118004) from the Ministry of Education, Culture, Sports, Science and Technology of Japan and by Grant-in-Aid for Scientific Research (B) (No. 25291035) from the Japan Society for the Promotion of Science.

Contents

1	**Introduction**	1
	References	4
2	**A New View on Mechanism of Functional Expression**	
	of an ATP-Driven Molecular Motor	5
	2.1 Coupling of a Molecular Motor and ATP Hydrolysis	
	Reaction.	5
	2.1.1 Thermodynamics of ATP Hydrolysis Reaction	5
	2.1.2 ATP Hydrolysis Cycle Where a Molecular Motor	
	Acts as a Catalyst for Hydrolysis Reaction	6
	2.2 Involvement of a Protein or Protein Complex Catalyzing	
	ATP Hydrolysis Reaction in ATP Hydrolysis Cycle	7
	2.3 Crucial Importance of Hydration Entropy in Functional	
	Expression of a Molecular Motor	8
	2.4 Mechanism of Force Generation by Water for Moving	
	or Rotating a Protein.	10
	2.5 Entropic Excluded-Volume Effect and Entropic Force	
	and Potential Generated by Water	11
	2.5.1 Entropy-Driven Formation of Ordered Structure	11
	2.5.2 Simple Examples of Entropic Force and Potential	14
	2.6 Roles of Electrostatic Interaction in Aqueous Solution Under	
	Physiological Condition	17
	2.7 Translational, Configurational Entropy of Water Leading	
	Receptor-Ligand Binding and Protein Folding	19
	2.8 Essential Roles of Water-Entropy Effect in Biological	
	Processes	22
	2.9 Recent Papers Pointing Out Crucial Importance of Hydration	
	Effect on Unidirectional Movement of Myosin Along F-Actin. . .	23
	2.10 Problems in Prevailing View on Functional Expression of a	
	Molecular Motor.	24

2.11 Inconsistency of Prevailing View with Some of Recent
        Experimental Facts   25
References   27

**3 Mechanism of Unidirectional Rotation of γ Subunit
    in F_1-ATPase**   29
    3.1 Definition of Packing Structure for a Protein or Protein
          Complex   29
    3.2 Nonuniform Binding of Nucleotides to $\alpha_3\beta_3$ or $\alpha_3\beta_3\gamma$
          Complex   30
    3.3 Theoretical Analyses on Packing Structure of $\alpha_3\beta_3\gamma$ Complex
          in Catalytic Dwell State   35
          3.3.1 A State of $\alpha_3\beta_3\gamma$ Complex Stabilized: Catalytic Dwell
                  State   35
          3.3.2 Methods of Theoretical Analyses   36
          3.3.3 Results of Theoretical Analyses   38
          3.3.4 Packing Structure Stabilized by Water-Entropy Effect . . .   40
          3.3.5 Relation Between Chemical Compound Bound
                  and Packing Efficiency in a β Subunit   42
    3.4 Normal Rotation Under Solution Condition that ATP
          Hydrolysis Reaction Occurs: Rotation Mechanism   42
          3.4.1 Basic Concept of Rotation Mechanism   42
          3.4.2 Details of Rotation Mechanism   44
          3.4.3 Crucial Importance of Water-Entropy Effect in
                  Unidirectional Rotation   47
          3.4.4 Change in System Free Energy During a Single
                  Rotation   49
          3.4.5 Effect of Electrostatic Attractive Interaction Between γ
                  and β Subunits   50
    3.5 Theoretical Analyses Based on Experimental Observations
          for Yeast F_1-ATPase   50
    3.6 Inverse Rotation Under Solution Condition that ATP Synthesis
          Reaction Occurs   52
          3.6.1 State of $\alpha_3\beta_3\gamma$ Complex Stabilized   52
          3.6.2 Details of Rotation Mechanism   53
    3.7 Normal and Inverse Rotations with the Same Frequency
          (Rotations in Random Directions) Under Solution Condition
          that ATP Hydrolysis and Synthesis Reactions Are
          Equilibrated   54
    3.8 Inverse Rotation Compelled by External Torque Imposed on
          Central Shaft and Occurrence of ATP Synthesis Under Solution
          Condition that ATP Hydrolysis Reaction Should Occur   54
          3.8.1 What Will Happen When Inverse Rotation is Forcibly
                  Executed?   54

3.8.2   Three Cases Where Normal Rotation Persists, Inverse
        Rotation Occurs, and Essentially no Rotations Occur
        When External Torque is Applied   56
3.8.3   Comparison with Experimental Results Observed When
        External Torque is Applied   58
3.8.4   Substantially Different Behavior Observed for a Mutant
        of F_1-ATPase   59
3.8.5 F_oF_1-ATP Synthase   61
References   61

4   **Concluding Remarks**   63
    4.1   Functional Rotation of AcrB   63
    4.2   Toward Investigation of Unidirectional Rotation of Central
          Shaft in V_1-ATPase   68
    References   68

5   **Appendix 1: Angle-Dependent Integral Equation Theory**   71
    References   74

6   **Appendix 2: Morphometric Approach**   77
    References   79

Chapter 1
Introduction

Abstract In the literature, we frequently come across such routine expressions as "an ATP-driven molecular motor converts chemical energy to mechanical work" and "a molecular motor realizes the reversible chemo-mechanical coupling". In the prevailing view, it is postulated that the molecular motor must perform mechanical work against the viscous resistance force by water during its functional expression such as the unidirectional movement and rotation. Namely, the molecular motor (a protein or protein complex) and the aqueous solution in which it is immersed are regarded as the system of interest and the external system, respectively. However, this regard is misleading. The system of interest comprises not only the molecular motor but also the aqueous solution, and the molecular motor performs essentially no mechanical work because the system volume does not change much and the system pressure is only 1 atm during its functional expression. We propose a new view pointing out that the force moving the protein or rotating a protein in the complex is generated by water. The entropic force originating from the translational displacement of water molecules in the whole system plays a pivotal role. As an example, we discuss the mechanism of unidirectional rotation of the central shaft (i.e., the γ subunit) in F_1-ATPase on the basis of the new view.

Keywords Molecular motor · ATP-driven protein · Chemo-mechanical coupling · Linear motor · Unidirectional movement · Rotary motor · Unidirectional rotation

There is an interesting class of molecular motors which functions by utilizing the ATP hydrolysis cycle comprising the ATP binding to the molecular motor, ATP hydrolysis (ATP + H_2O → ADP + Pi), and dissociation of ADP and Pi from the molecular motor. Typical examples are a linear molecular motor, myosin, which performs unidirectional movement along filamentous actin (F-actin) and a rotary molecular motor, F_1-ATPase, in which its central shaft (i.e., the γ subunit) performs unidirectional rotation. The molecular motor is often referred to as the "ATP-driven protein or protein complex". Unfortunately, the mechanism of its functional expression (e.g., the unidirectional movement and rotation) remains quite elusive.

© The Author(s), under exclusive license to Springer Nature Singapore Pte Ltd. 2021   1
M. Kinoshita, *Mechanism of Functional Expression of F_1-ATPase*,
SpringerBriefs in Molecular Science,
https://doi.org/10.1007/978-981-33-6232-1_1

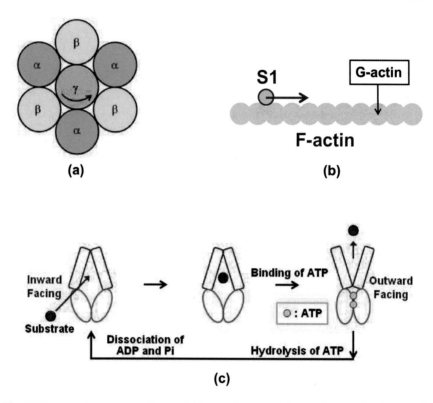

Fig. 1.1 Functional expression of an ATP-driven protein or protein complex (a molecular motor) utilizing hydrolysis cycle comprising ATP binding to the protein or protein complex, ATP hydrolysis, and dissociation of ADP and Pi from the protein or protein complex. **a** Unidirectional rotation of γ subunit in $\alpha_3\beta_3\gamma$ complex of F_1-ATPase. The ATP hydrolysis cycle occurs for the $\alpha_3\beta_3$ complex. **b** Unidirectional movement of myosin subfragment 1 (S1) along filamentous actin (F-actin). In the experiment by Kitamura et al. [3, 4], S1 is forcibly attached to F-actin using a novel technique. The ATP hydrolysis cycle occurs for S1. **c** Transport of a substrates across membrane by ATP-binding cassette (ABC) transporter. The transporter takes the inward-facing structure for the substrate insertion, while it takes the outward-facing structure for the substrate release. The ATP hydrolysis cycle occurs for the transporter

In protein folding, for example, the structures of a protein in the unfolded and folded states (i.e., those before and after the folding) are different. The folded structure (in a strict sense, the structural ensemble of folded structures) is stabilized in the equilibrium state where the free energy of the protein-water system is minimized. It can readily be understood that the folding is an irreversible process accompanied by a decrease in system free energy (i.e., a spontaneously occurring process). On the other hand, the structures of a molecular motor before and after its functional expression are the same. As illustrated in Fig. 1.1, before and after the 120° rotation of the γ subunit incorporated in the $\alpha_3\beta_3$ complex (F_1-ATPase is the $\alpha_3\beta_3\gamma$ complex) [1, 2], before and after the 2.5-step (on an average) movement of myosin subfragment 1 (S1) along

F-actin [3, 4], and before the substrate insertion into ATP-binding cassette (ABC) transporter [5, 6] and after the substrate release from it, respectively, the structures of the $\alpha_3\beta_3\gamma$ complex, S1, and transporter are the same. (In this book, ABC transporter is also considered as an ATP-driven molecular motor.) Therefore, it appears that the stabilized structure (or the equilibrium state) must be destructed for the functional expression. One might be inclined to think that energy or free energy is required for the destruction. In the prevailing view, the following explanations are made: The chemical energy stored in an ATP molecule or the free energy of ATP hydrolysis reaction (i.e., energy or free-energy supply by ATP) is utilized for the destruction; myosin or the γ subunit must perform mechanical work against the viscous resistance force by water for achieving the unidirectional movement or rotation [7, 8], and the energy or free-energy supplied by ATP is converted to this work; and a molecular motor is characterized by its ability of reversible chemo-mechanical coupling implying that it can transduce the chemical energy or free energy to the mechanical work and vice versa. In the prevailing view, the molecular motor and the aqueous solution in which it is immersed are regarded as the system of interest and the external system, respectively.

In this book, by pointing out that the prevailing view is physically problematic and conflicting with some of recent experimental observations, we suggest a completely new view. Its outline is as follows. The ATP hydrolysis reaction, an irreversible process, is catalyzed by the molecular motor. The molecular motor is thus involved in the ATP hydrolysis cycle referred to above. The functional expression of the molecular motor is accompanied by a decrease in system free energy and a spontaneously occurring process. The system of interest comprises not only the molecular motor but also the aqueous solution, and it performs essentially no mechanical work because the system volume does not change much and the system pressure is only 1 atm during the ATP hydrolysis cycle (see Sect. 2.10 for more details). The force which moves myosin or rotates the γ subunit is generated by nothing but water. Especially, the entropic force ascribed to the translational displacement of water molecules in the whole system plays a pivotal role. The concept of reversible chemo-mechanical coupling is physically irrelevant. Interestingly, the functional expression of a molecular motor is hindered by water in the prevailing view, whereas it is driven by water in the new view. Choosing the unidirectional rotation of the γ subunit in F_1-ATPase as a paradigmatic example, we discuss its mechanism in detail.

The molecular motors were already considered in our earlier books [9, 10]. In the first book [9], we summarized the water roles in functional expression of the molecular motors as an extension of biological self-assembly and ordering processes with which we had been dealing (refer to our review articles [11–14]). In the second book [10], by choosing actomyosin (i.e., myosin and F-actin), we raised questions about the prevailing view on the mechanism of functional expression of the molecular motors. In this book, we explain how the prevailing view is problematic and inconsistent with some of recent experimental observations in much more detail, and construct a new view in a more complete and convincing form. Moreover, taking the unidirectional rotation of the γ subunit in F_1-ATPase as a paradigmatic example, we show, for the first time, that the following four scenarios can be elucidated in a unified manner within the same theoretical framework: (A) rotation in the normal direction under the solution condition that the ATP hydrolysis reaction occurs; (B) rotation in the inverse direction under the solution condition that the ATP synthesis reaction occurs; (C) rotations in random directions under the solution condition that the ATP hydrolysis and synthesis reactions are in equilibrium; and (D) occurrence of ATP synthesis, even under the solution condition that the ATP hydrolysis reaction should occur, through forcible rotation in the inverse direction by means of sufficiently strong external torque imposed on the γ subunit. Scenarios (B) and (C) have not yet been corroborated in experiments but should actually happen in our view. F_1-ATPase was considered in our two papers [15, 16] and the first book [9], but only scenario (A) was treated, and besides, just in a preliminary form. Many of the ideas described in this book have newly been developed by the author.

References

1. Shimabukuro K, Yasuda R, Muneyuki E, Hara KY, Kinosita K Jr, Yoshida M (2003) Proc Natl Acad Sci USA 100:14731
2. Adachi K, Oiwa K, Nishizaka T, Furuike S, Noji H, Itoh H, Yoshida M, Kinosita K Jr (2007) Cell 130:309
3. Kitamura K, Tokunaga M, Iwane AH, Yanagida T (1999) Nature 397:129
4. Kitamura K, Tokunaga M, Esaki S, Iwane AH, Yanagida T (2005) Biophysics 1:1
5. Ward A, Reyes CL, Yu J, Roth CB, Chang G (2007) Proc Natl Acad Sci USA 104:19005
6. Hollenstein K, Dawson RJP, Locher KP (2007) Curr Opin Struct Biol 17:412
7. Noji H, Yasuda R, Yoshida M, Kinosita K Jr (1997) Nature 386:299
8. Ananthakrishnan R, Ehrlichter A (2007) Int J Biol Sci 3:303
9. Kinoshita M (2016) Mechanism of functional expression of the molecular machines. Springer Briefs in Molecular Science, Springer, ISBN: 978-981-10-1484-0
10. Kinoshita M (2018) Functioning mechanism of ATP-driven proteins inferred on the basis of water-entropy effect. In: Suzuki M (ed) The role of water in ATP hydrolysis energy transduction by protein machinery. Chapter 18, Part III, Springer Briefs in Molecular Science, Springer, ISBN: 978-981-10-8458-4, pp 303–323
11. Kinoshita M (2003) Surface-induced phase transition and long-range surface force: roles in colloidal and biological systems. Transworld Research Network, India, ISBN: 81-7895-113-4, Vol. 1, Recent Research Developments in Molecular Physics, pp 21–41
12. Kinoshita M (2009) Front Biosci 14:3419
13. Kinoshita M (2009) Int J Mol Sci 10:1064
14. Kinoshita M (2013) Biophys Rev 5:283
15. Yoshidome T, Ito Y, Ikeguchi M, Kinoshita M (2011) J Am Chem Soc 133:4030
16. Yoshidome T, Ito Y, Matubayasi N, Ikeguchi M, Kinoshita M (2012) J Chem Phys 137:035102

Chapter 2
A New View on Mechanism of Functional Expression of an ATP-Driven Molecular Motor

Abstract The ATP hydrolysis reaction can hardly occur in bulk aqueous solution without a catalyst. An ATP-driven molecular motor, which acts as the catalyst, is coupled with the reaction. Consequently, the reaction proceeds as the ATP hydrolysis cycle comprising the following events: the ATP binding to the molecular motor, hydrolysis of ATP into ADP and Pi, and dissociation of ADP and Pi from the molecular motor. Water in which ATP, ADP, and Pi are dissolved as well as the molecular motor (a protein or protein complex) forms the system of interest. In this system, the ATP hydrolysis reaction catalyzed by the molecular motor occurs as an irreversible process (i.e., a spontaneously occurring process). Upon each event in the ATP hydrolysis cycle, the molecular motor exhibits sequential changes in its configuration for minimizing the system free energy. We rationalize the argument that the translational, configurational entropy of water is a principal component of the system free energy. A protein (e.g., myosin) is moved or a protein in the complex (e.g., the γ subunit in the $\alpha_3\beta_3\gamma$ complex of F_1-ATPase) is rotated in the direction where the water entropy already maximized can be retained.

Keywords Self-assembly · Protein folding · Molecular recognition · Hydrophobic effect · ATP · ADP · Actomyosin

2.1 Coupling of a Molecular Motor and ATP Hydrolysis Reaction

2.1.1 Thermodynamics of ATP Hydrolysis Reaction

In the prevailing view, the attention is paid to the chemical energy stored in an ATP molecule or the free energy of ATP hydrolysis reaction. However, the important quantity is the free energy of ATP hydrolysis reaction.

The change in standard free energy $\Delta G° < 0$ of the chemical reaction, ATP + H_2O → ADP + Pi (this is written as "$ATP^{4-}+H_2O \rightarrow ADP^{3-}+Pi^{2-}+H^+$" when the charge balance is emphasized), is ~$-12k_BT$ ($T = 298$ K) [1] where k_B is the Boltzmann

© The Author(s), under exclusive license to Springer Nature Singapore Pte Ltd. 2021   5
M. Kinoshita, *Mechanism of Functional Expression of F_1-ATPase*,
SpringerBriefs in Molecular Science,
https://doi.org/10.1007/978-981-33-6232-1_2

constant and T is the absolute temperature ($k_{\mathrm{B}}T = 0.592$ kcal/mol at $T = 298$ K). $\Delta G°$ is "the formation free energy of 1 mol of ADP plus that of 1 mol of Pi" minus "the formation free energy of 1 mol of ATP plus that of 1 mol of H_2O" in the standard state where the concentrations of ATP, ADP, and Pi equal 1 mol/L, $T = 298$ K, and the pressure P is 1 atm. The free-energy change upon the chemical reaction ΔG is approximately related to $\Delta G°$ through

$$\Delta G = \Delta G° + RT\ln(Z), \; Z = [\text{ADP}][\text{Pi}]\big/[\text{ATP}] \tag{2.1}$$

where R is the gas constant and $[X]$ is the dimensionless concentration of X (i.e., the concentration of X in mol/L divided by 1 mol/L) in the aqueous solution. The chemical potential of water in the aqueous solution is assumed to be equal to that of pure water. A value of Z is valid only for given values of pH and $[\text{Mg}^{2+}]$ [2, 3].

When [ATP] is sufficiently high and [ADP] and [Pi] are sufficiently low such that $\ln(Z)$ takes a negative or small, positive value and $\Delta G < 0$, the reaction frequency in the right direction (ATP hydrolysis: ATP + H_2O → ADP + Pi) is much higher than that in the left direction (ATP synthesis: ADP + Pi → ATP + H_2O). That is, the overall reaction under this solution condition is the ATP hydrolysis. ΔG is dependent on the concentrations of ATP, ADP, and Pi but roughly equal to $-20k_{\mathrm{B}}T$ ($T = 298$ K) in aqueous solution under the physiological condition [4]. When [ATP] is sufficiently low and [ADP] and [Pi] are sufficiently high such that $\ln(Z)$ is positive and $|\Delta G°|$ $< RT\ln(Z)$, $\Delta G > 0$. Under this solution condition, the overall reaction is the ATP synthesis. In the chemical equilibrium state, $\Delta G = 0$ and the reaction frequencies in the right and left directions are the same.

2.1.2 ATP Hydrolysis Cycle Where a Molecular Motor Acts as a Catalyst for Hydrolysis Reaction

Without a catalyst, the reaction rate of the chemical reaction, ATP + H_2O → ADP + Pi, in bulk aqueous solution is extremely low. Importantly, an ATP-driven molecular motor acts as the catalyst. In other words, the molecular motor is coupled with the reaction. The reaction proceeds as the ATP hydrolysis cycle comprising the following three events: (1) ATP binding to the molecular motor, (2) hydrolysis of ATP into ADP and Pi, and (3) dissociation of ADP and Pi from the molecular motor.

As explained in Fig. 2.1, the binding free energy of A (A is ATP or ADP) and the molecular motor M, $\Delta G_{\mathrm{B}}°(\text{A})$, is "the free energy of 1 mol of MA" minus "the free energy of 1 mol of M plus that of 1 mol of A" in the standard state. $\Delta G_{\mathrm{B}}°(\text{ATP})$ and $\Delta G_{\mathrm{B}}°(\text{ADP})$ are both negative. Let $\Delta G_{\mathrm{B}}(\text{A})$ be the free-energy change upon the binding of A to M. Under the solution condition that [ATP] and [ADP] are sufficiently high and low, respectively, $\Delta G_{\mathrm{B}}(\text{ATP}) < 0$ and $\Delta G_{\mathrm{B}}(\text{ADP}) > 0$: ATP binds to M (M + ATP → MATP) and ADP dissociates from M (MADP → M + ADP). The discussion to be made for Pi is similar to that for ADP.

M: Molecular motor, A: ATP or ADP.
Binding process: M+A→MA.

Binding free energy (free-energy change occurring when 1 mol of MA is formed from 1 mol of M and 1 mol of A):

$$\Delta G_B = \mu_{MA} - \mu_M - \mu_A,$$

$$\mu_M = \mu_M° + RT\ln(C_M/C_M°), \ \mu_A = \mu_A° + RT\ln(C_A/C_A°), \ \mu_{MA} = \mu_{MA}° + RT\ln(C_{MA}/C_{MA}°).$$

$$C_M°, C_A°, C_{MA}°: 1 \text{ mol/L.}$$

$$\Delta G_B = \mu_{MA}° - \mu_M° - \mu_A° + RT\ln\{C^*_{MA}/(C^*_M C^*_A)\},$$

$$C^*_M = C_M/C_M°, \ C^*_A = C_A/C_A°, \ C^*_{MA} = C_{MA}/C_{MA}°.$$

$$\Delta G_B = \Delta G_B° + RT\ln\{C^*_{MA}/(C^*_M C^*_A)\}, \ \Delta G_B° = \mu_{MA}° - \mu_M° - \mu_A°.$$

Even when $\Delta G_B°$ is negative, ΔG_B becomes positive if C^*_A is sufficiently low:
A dissociates from M (MA→M+A).

Fig. 2.1 Thermodynamics of nucleotide binding to or dissociation of nucleotide from a molecular motor M. A is ATP or ADP, MA is M to which A is bound, μ_J= chemical potential of J (J = M, A, and MA), and C_J= concentration of J. The superscript "°" denotes the standard state where the concentrations of M, A, and MA equal 1 mol/L, $T = 298$ K, and $P = 1$ atm. $C^*_J = [J] = $ dimensionless concentration of J. When the solution is under the condition that the ATP and ADP concentrations are sufficiently high and low, respectively, ATP binds to and ADP dissociates from M

The molecular motor, whose affinity for ATP is higher than that for ADP ($\Delta G_B°(\text{ATP}) < \Delta G_B°(\text{ADP}) < 0$), functions under the solution condition that [ATP] is kept sufficiently high and [ADP] and [Pi] are kept sufficiently low. Each of the ATP binding to the molecular motor (event (1)), ATP hydrolysis (event (2)), and dissociation of ADP and Pi from the molecular motor (event (3)) is accompanied by a decrease in system free energy and spontaneously occurs.

2.2 Involvement of a Protein or Protein Complex Catalyzing ATP Hydrolysis Reaction in ATP Hydrolysis Cycle

In this section, we consider a protein or protein complex catalyzing the ATP hydrolysis reaction: myosin, ABC transporter, and $\alpha_3\beta_3$ complex in F_1-ATPase. It is assumed that the protein or protein complex is isolated. Under the solution condition that the ATP concentration is sufficiently high and the ADP and Pi concentrations are sufficiently low, upon each of events (1), (2), and (3) in the ATP hydrolysis cycle (see Sect. 2.1.2), the protein or protein complex exhibits a structural change as depicted in Fig. 2.2. We note that in the absence of ATP, ADP, and Pi, structure I would remain

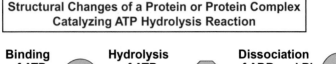

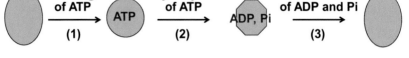

Fig. 2.2 Structural changes of a protein (e.g., myosin) or protein complex (e.g., ABC transporter and the $\alpha_3\beta_3$ complex in F_1-ATPase) catalyzing ATP hydrolysis reaction. The reaction can be accelerated by not an isolated β subunit but a β subunit in the complex (the arginine finger in the α subunit plays an imperative role: see Ref. 36 in Chap. 3). In the aqueous solution, the ATP concentration is sufficiently high, and the ADP and Pi concentrations are sufficiently low. The ATP hydrolysis cycle comprises events (1), (2), and (3) where the ATP binding to the protein or protein complex, ATP hydrolysis, and dissociation of ADP and Pi from the protein or protein complex take place, respectively. The protein or protein complex exhibits a structural change upon each of the three events. In particular, the structural changes upon events (1) and (3) are substantially large

stabilized and unchanged. Each event occurs as an irreversible process accompanied by a decrease in system free energy. This cycle is repeated. The decrease in system free energy upon the ATP binding or the ADP dissociation can be justified as discussed in Sect. 2.1.2. The net decrease after a single cycle where one ATP molecule is hydrolyzed is ΔG. (Two ATP molecules are hydrolyzed in ABC transporter: The net decrease is $2\Delta G$ if the increase in system free energy originating from the active transport of a substrate against the substrate concentration gradient is not taken into account.) As mentioned above, ΔG is roughly equal to $-20k_BT$ ($T = 298$ K) [4] (see Sect. 2.1.1). Let $\Delta G_{(I)}$ ($I = 1, 2, 3$) be the free-energy decrease upon event (I). Since $\Delta G_{(1)} + \Delta G_{(2)} + \Delta G_{(3)} = \Delta G$, the absolute value of free energy of ATP hydrolysis in the protein or protein complex, $|\Delta G_{(2)}|$, is considerably smaller than $|\Delta G|$.

2.3   Crucial Importance of Hydration Entropy in Functional Expression of a Molecular Motor

When a protein or protein complex catalyzing the ATP hydrolysis reaction (see Sect. 2.2) functions, another protein, substrate, or protein complex coexists with it. As illustrated in Fig. 2.3, we consider three systems comprising solutes 1 and 2 immersed in aqueous solution of ATP, ADP, and Pi. It is imperative to account for the hydration of the two solutes. The hydration of the protein or a protein in the complex is substantially influenced by the aforementioned coexistence especially in systems (I) and (III).

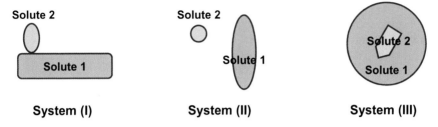

Fig. 2.3 Three representative systems considered. System (I) (actomyosin): F-actin (solute 1) and myosin (solute 2). System (II): ABC transporter (solute 1) and a small solute (solute 2). System (III) (F_1-ATPase): $\alpha_3\beta_3$ complex (solute 1) and γ subunit (solute 2). Solutes 1 and 2 are immersed in aqueous solution of ATP, ADP, and Pi. In the aqueous solution, the ATP concentration is sufficiently high, and the ADP and Pi concentrations are sufficiently low

System (I). Solute 1 is F-actin and solute 2 is myosin (or S1) [5, 6]. Myosin corresponds to the protein catalyzing the ATP hydrolysis reaction. The position of solute 1 is fixed whereas that of solute 2 is variable. The system free energy F is strongly dependent on the structures of solutes 1 and 2 and the position of solute 2. In the absence of ATP, ADP, and Pi, the structures of solutes 1 and 2 and the position of solute 2 remain unchanged once they are stabilized for minimizing F. In the system of interest, however, the ATP hydrolysis cycle comes into play. Upon each event in this cycle (see Sect. 2.1.2), solute 2 undergoes a structural change. (The structure of solute 1 also changes.) As a result, F does not take the lowest value for each new system configuration and solute 2 is moved (i.e., the position of solute 2 is changed) so that F can be minimized. More strictly, upon each event in the ATP hydrolysis cycle, the structures of solutes 1 and 2 as well as the position of solute 2 are changed for minimizing F.

System (II). Solute 1 is ABC transporter and solute 2 is a substrate [7, 8]. The transporter corresponds to the protein complex catalyzing the ATP hydrolysis reaction. The position of solute 1 is fixed whereas that of solute 2 is variable. The system free energy F is strongly dependent on the structure of solute 1 and the position of solute 2. Upon each event in the ATP hydrolysis cycle, solute 1 undergoes a structural change. As a result, F does not take the lowest value for each new system configuration and solute 2 is moved (i.e., the position of solute 2 is changed) for minimizing F (see the release of the substrate illustrated in Fig. 1.1c).

System (III). Solute 1 is the $\alpha_3\beta_3$ complex and solute 2 is the γ subunit in the $\alpha_3\beta_3\gamma$ complex of F_1-ATPase [9, 10]. The $\alpha_3\beta_3$ complex corresponds to the protein complex catalyzing the ATP hydrolysis reaction. The positions of solutes 1 and 2 are fixed and the orientation of solute 2 is variable. The system free energy F is strongly dependent on the structure of solute 1 and the orientation of solute 2. Upon each event in the ATP hydrolysis cycle, solute 1 undergoes a structural change. As a result, F does not take the lowest value for each new system configuration and solute 2 is rotated (i.e., the orientation of solute 2 is changed) so that F can be minimized.

The ATP hydrolysis cycle is repeated as an irreversible process during which myosin performs unidirectional movement along F-actin, diverse substrates are transported across the membrane by ABC transporter, and the γ subunit performs unidirectional rotation in F_1-ATPase. In cases of actomyosin and F_1-ATPase, the cycle is accompanied by a decrease in system free energy of $\Delta G \sim -20k_B T$ ($T = 298$ K).

F is composed of the conformational (intramolecular) energy, conformational entropy, and hydration free energy of solutes 1 and 2. The hydration free energy is the sum of the hydration energy and entropy. The absolute value of hydration entropy of a solute is the magnitude of water-entropy loss caused by the insertion of the solute. In general, a change in conformational energy and that in hydration energy are compensating. That is, the latter is positive when the former is negative and the latter is negative when the former is positive. The sum of the two energies remains roughly constant, i.e., the change in this sum is rather small. (In a strict sense, it is often that the change in the hydration energy is larger.) On the other hand, a conformational-entropy change is significantly smaller than the hydration-entropy change. (See Sects. 2.6–2.8 for more details.) Therefore, the hydration entropy can be treated as a principal component of F. It follows that F and the water entropy S_{Water} is approximately related through

$$F \sim -T S_{\text{Water}}. \tag{2.2}$$

In system (I) shown in Fig. 2.3, for example, we can state that the structures of the two solutes and the position of solute 2 are determined so that the water entropy can be maximized.

2.4 Mechanism of Force Generation by Water for Moving or Rotating a Protein

Let us consider solutes 1 and 2 in system (I) (see Fig. 2.3). We assume that the structures of the two solutes are fixed for simplicity. F is a function of the Cartesian coordinates of the center of gravity of solute 2, (x, y, z). The origin of the coordinate system $(0, 0, 0)$ is taken to be, for example, the left edge of solute 1. "$F(x, y, z) - F(+\infty, +\infty, +\infty)$" represents the spatial distribution of the potential of mean force (PMF: the water-mediated interaction) between solutes 1 and 2. The mean force acting on solute 2, f, is expressed as

$$\boldsymbol{f} = f_x \mathbf{i} + f_y \mathbf{j} + f_z \mathbf{k}, f_x = -\partial F / \partial x, \ f_y = -\partial F / \partial y, \ f_z = -\partial F / \partial z \tag{2.3}$$

where $\mathbf{i}$, $\mathbf{j}$, and $\mathbf{k}$ denote the direction unit vectors. $f(x_0, y_0, z_0)$ represents the force induced between solutes 1 and 2 averaged over all the possible configurations of water molecules in the entire system with (x, y, z) being fixed at (x_0, y_0, z_0). We can take the view that a potential or force field acts on solute 2 near solute 1 (or equivalently, solute 2 feels a potential or force field due to the presence of solute 1 near it) [11]. Since S_{Water} in Eq. (2.2) is quite large and strongly dependent on (x, y, z), the entropic force expressed by Eqs. (2.2) and (2.3) plays substantial roles.

When the structures of solutes 1 and 2 are made variable as in the real system, the PMF should shift downward (i.e., in the negative direction) especially in regions where the PMF is positive and large. This is because the solutes always exhibit structural changes so that F can become as low as possible.

We then consider solutes 1 and 2 in system (III) (see Fig. 2.3). We assume that the structures of the two solutes are fixed for simplicity. F is a function of the rotation angle θ defined for solute 2. The mean torque acting on solute 2, $\tau(\theta)$, is expresses as

$$\tau = -\partial F / \partial \theta. \tag{2.4}$$

$\tau(\theta_0)$ represents the torque acting on solute 2 averaged over all the possible configurations of water molecules in the entire system with θ being fixed at θ_0. In this case, F becomes lowest at $\theta = \theta_{\min}$ and $\tau(\theta_{\min}) = 0$. Since S_{Water} in Eq. (2.2) is quite large and strongly dependent on θ, the entropic torque expressed by Eqs. (2.2) and (2.4) plays substantial roles.

2.5   Entropic Excluded-Volume Effect and Entropic Force and Potential Generated by Water

2.5.1   Entropy-Driven Formation of Ordered Structure

The concept referred to as the "entropic excluded-volume effect" or the "solvent-entropy effect" ("water-entropy effect" when the solvent is water) is crucially important in colloidal and biological systems [11, 12]. As illustrated in Fig. 2.4a, the insertion of a solute into water causes the generation of a space which is inaccessible to the centers of water molecules (the space occupied by the solute itself plus the space shown in gray). If water molecules are spheres with diameter d_S and the solute is a sphere with diameter d_L, the excluded space is a sphere with diameter "$d_S + d_L$". The volume of the excluded space is the "excluded volume (EV)". Upon the contact of two solutes (see Fig. 2.4b), the two excluded spaces overlap, the total EV reduces by the volume of the overlapping space marked in black, and the total volume available for the translational displacement of water molecules increases by the same volume. The contact leads to an increase in the number of accessible translational configurations of water molecules (i.e., the number of possible coordinates of the centers of water molecules). This increase is followed by a gain of the configurational entropy of water. An interaction driving the solutes to contact each other is thus induced [11, 12]. Suppose that the solutes and water molecules are

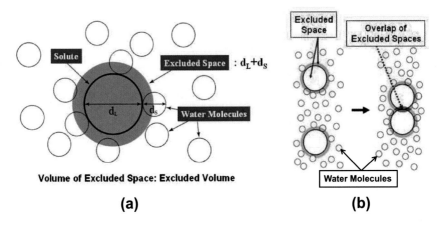

Fig. 2.4 **a** A solute immersed in water. Water molecules are spheres with diameter d_S and the solute is a sphere with diameter d_L. The excluded space is inaccessible to the centers of water molecules. **b** Contact of two solutes in water. The two excluded spaces overlap, the total excluded volume reduces by the volume of the overlapping space marked in black, and the total volume available for the translational displacement of water molecules increases by the same volume, leading to a gain of the configurational entropy of water

modeled as neutral hard spheres interacting through hard-body potentials including no soft attractive and repulsive potentials (of course, we usually adopt realistic solute and water models in our theoretical analyses). Even in this model system, where all the allowed system configurations (i.e., configurations without the overlap of hard bodies) share the same energy and the system behavior is purely entropic in origin, the aforementioned interaction is induced. Therefore, the induced interaction is called the "entropic potential". Strictly, the entropic potential is dependent not only on the volume of overlapping space marked in black but also on the microscopic structure of the water molecules confined by the two solute surfaces.

Let r be the distance between the centers of two spherical solutes. The entropic force $f_{ent}(r)$ induced between these solutes is related to the entropic potential $u_{ent}(r)$ by $f_{ent} = -du_{ent}/dr$. $u_{ent}(r)$ represents the entropic component of the PMF. $u_{ent}(r_0)/(-T)$ represents the entropy of water for $r = r_0$ relative to that for $r \to \infty$. $f_{ent}(r_0)$ represents the entropic force induced between the solutes averaged over all the possible configurations of water molecules in the whole system with r being fixed at r_0.

It is very interesting to consider highly nonspherical solutes [13]. Figure 2.5 illustrates four different manners for the contact of solutes with cylindrical or disc-like shapes. The volume of the overlapping excluded spaces is maximized in manner 4, the most ordered contact, leading to the largest gain of the translational, configurational entropy of water upon the contact. Water thus drives the solutes to contact each other in manner 4. Though such a solute contact causes a decrease in the translational, configurational entropy of solutes themselves, the water-entropy increase dominates and the system entropy increases unless the solute concentration is extremely low.

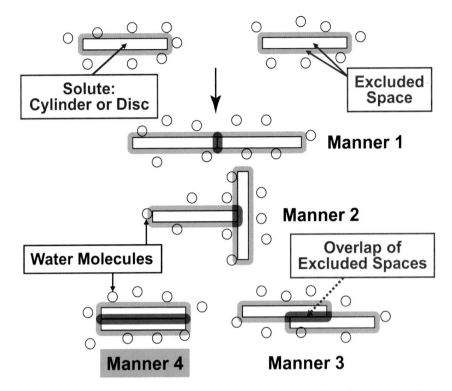

Fig. 2.5 Four different manners of contact of two cylindrical or disc-like solutes in water. Water drives the solutes to contact each other in manner 4, the most ordered contact, which leads to the largest gain of the translational, configurational entropy of water upon the contact

This is the physical essence of the concept of "entropically driven self-assembly process" proposed by us [14]. Solutes immersed in water are driven to form an ordered structure to increase the system entropy.

The "entropic EV effect" or "solvent-entropy effect" [11, 12, 14, 15] described above becomes stronger with increasing η_S or decreasing d_S. Here, η_S is the packing fraction of the solvent defined as $\eta_S = \pi \rho_S d_S^3/6$ where ρ_S and d_S are, respectively, the number density and the molecular diameter of the solvent. The free-energy decrease occurring when a pair of spherical solutes with diameter d_L contact each other can be approximated by $-1.5k_B T \eta_S (d_L/d_S)$ for $d_L/d_S \gg 1$. By virtue of the hydrogen-bonding network, water can exist as a dense liquid at ambient temperature and pressure despite its exceptionally small molecular diameter. Among the ordinary liquids in nature, the entropic EV effect becomes strongest for water [16]. Though neon and water share almost the same value of d_S, neon is in gas state at ambient temperature and pressure, resulting in a negligibly small entropic EV effect when the solvent is neon. Since the molecular diameter of cyclohexane is considerably larger than that of water, cyclohexane presents only a significantly weaker entropic EV effect though it exists as a dense liquid. For $\eta_S = 0.383$ (this is the value for water at 298 K and 1 atm) and $d_L/d_S = 20$, for example, $-1.5k_B T\eta_S(d_L/d_S)$ reaches an unexpectedly large value, $-11.5k_B T$. In a biological system where large biomolecules such as proteins or protein complexes are immersed in water, the entropic EV effect is remarkably large.

2.5.2 Simple Examples of Entropic Force and Potential

In Fig. 2.6, we show examples of the entropic force $F_{Wall}(h)$ and potential $\Phi_{Wall}(h)$ $(F_{Wall}(h) = -d\Phi_{Wall}(h)/dh)$ induced between a large sphere and a planar wall immersed in small spheres (h is the nearest distance between large-sphere and wall surfaces). The spheres are all neutral hard spheres and the wall is a neutral hard wall. In this hard-body model system, all the allowed system configurations share the same energy and the system behavior is purely entropic in origin. $F_{Wall}(h)$ and $\Phi_{Wall}(h)$

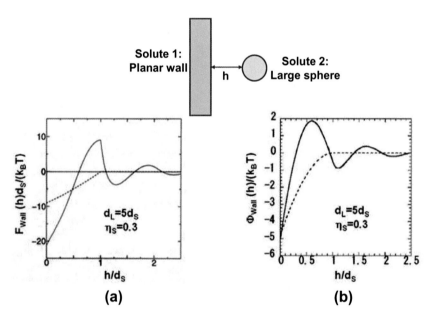

Fig. 2.6 Entropic force (**a**) and potential (**b**) induced between a large sphere and a planer wall immersed in small spheres. The spheres are all neutral hard spheres and the wall is a neutral hard wall. The position of the wall is fixed. The broken and solid lines denote the curves calculated by the Asakura-Oosawa theory and by an elaborate statistical-mechanical theory, respectively. The solid line can be considered to be exact. T, k_B, h, d_L, and d_S denote the absolute temperature, Boltzmann constant, surface separation, diameter of the large sphere, and diameter of the small spheres, respectively, and η_S is the packing fraction of the small spheres. In (**b**), once the large sphere contacts the wall, the large sphere must overcome a free-energy barrier of ~$7k_B T$ to get detached from the wall. In (**a**) and (**b**), d_L is set at $5d_S$. The amplitudes of the solid lines become larger with an increase in η_S or d_L

are, respectively, the mean force and the PMF possessing no energetic components. $\Phi_{\text{Wall}}(h_0)$ represents the free energy of small spheres for $h = h_0$ relative to that for $h \to \infty$.

The curves of $F_{\text{Wall}}(h)$ and $\Phi_{\text{Wall}}(h)$ calculated using the Asakura-Oosawa (AO) theory [17, 18] and an elaborate statistical-mechanical theory (e.g., the integral equation theory [19]) are compared in Fig. 2.6. In the AO theory, the force and potential are evaluated using only the volume of overlapping excluded space (see Fig. 2.7). No force or potential is induced between the large sphere and the wall when there is no overlap of the two excluded spaces generated by the large sphere and the wall. The overlap occurs only for $h < d_S$ and a force is induced to increase the overlapped volume and the entropy of small spheres. The force, which is always attractive, becomes monotonically stronger with a decrease in h. On the other hand, the exact force is oscillatory with a periodicity of d_S and longer ranged. This complex behavior is attributed to the microscopic structure of the small spheres formed within the domain confined by the large sphere and the wall. This microscopic structure is neglected in the AO theory. It is observed in Fig. 2.6 that the two curves share almost the same value of $\Phi_{\text{Wall}}(0)$. We note that in cases where a large sphere contacts

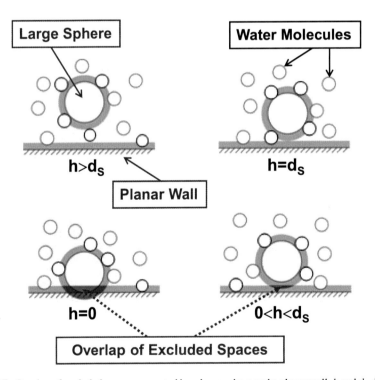

Fig. 2.7 Overlap of excluded spaces generated by a large sphere and a planar wall. h and d_S denote the surface separation and the diameter of the small spheres, respectively. The overlapping space is marked in black. This figure is prepared for explaining the Asakura-Oosawa theory [17, 18] in which the force and potential are evaluated using only the volume of overlapping excluded space

another large sphere or a planar wall, fortuitous cancellation of errors occurs in the calculation of $\Phi_{\text{Wall}}(0)$ using the AO theory, leading to the almost exact value of $\Phi_{\text{Wall}}(0)$ [11]. (Such cancellation does not occur for the contact of solutes with more complex geometric properties [11]).

Here, we discuss the physical origin of the oscillatory behavior of the entropic potential $\Phi_{\text{Wall}}(h)$, the solid line shown in Fig. 2.6b [20]. When h is close to nd_S ($n = 1, 2, \ldots$), $\Phi_{\text{Wall}}(h)$ takes a negative, local-minimum value. We previously showed that as d_L increases, the local-minimum positions become closer to nd_S. On the other hand, the surface separation h which is not close to nd_S ($n = 1, 2, \ldots$) is entropically unfavorable with the result of positive $\Phi_{\text{Wall}}(h)$. These results can be rationalized as follows [20]. Importantly, the presence of a small sphere also generates an EV for the other small spheres. In this sense, all the small spheres in the system are entropically correlated and this entropic correlation is referred to as the "crowding of small spheres". The entropy of small spheres decreases as the crowding becomes more significant. The small spheres are driven to be packed within the space confined between the large sphere and the wall for increasing the overlap of EVs generated by the large sphere, wall, and small spheres, increasing the total volume available for the translational displacement of small spheres in the system, and reducing the crowding of small spheres. The small spheres within the confined domain are entropically unfavorable, but the effect of the reduced crowding dominates. Close packing of the small spheres within the confined domain is achieved for $h \sim nd_S$. Hence, the entropy of small spheres for $h \sim nd_S$ becomes higher than that for $h \to \infty$, leading to a negative, local-minimum value of $\Phi_{\text{Wall}}(h)$. When $h \sim nd_S$ does not hold, it is clear from simple geometric consideration that a significant amount of void space is inevitably formed within the confined domain. This formation causes a decrease in the total volume available for the translational displacement of small spheres in the system. As a consequence, the entropy of small spheres becomes lower than that for $h \to \infty$, giving rise to positive $\Phi_{\text{Wall}}(h)$.

When the small neutral hard spheres are replaced by water molecules, the energetic factor as well as the entropic one comes into play. For the energetic factor, the induced interaction is considerably more influenced by the properties of large-sphere and wall surfaces. The mean force and the PMF then exhibit downward shifts with smaller amplitudes of the oscillatory curves [21]. This can readily be understood because the large-sphere and wall surfaces, which cannot form hydrogen bonds with water molecules, become highly unfavorable, driving their contact more strongly for reducing the area of the surfaces exposed to water. Suppose that the large neutral hard sphere and the neutral hard wall are replaced by a large sphere and a wall possessing the surfaces comprising atoms with positive and negative partial charges, respectively. Due to the surface-water electrostatic attractive interactions, the large-sphere and wall surfaces become energetically more affinitive for water. This factor hinders their contact because a larger area of the surfaces exposed to water is more favored. The mean force and the PMF then exhibit upward shifts with larger amplitudes of the oscillatory curves [21]. Consequently, the resultant mean force and PMF look more like the entropic force and potential for the hard-body model, respectively. In the PMF between two biomolecules, its entropic component usually dominates. It is worthwhile to note that η_S of water is 0.383, much higher than the value set in Fig. 2.6, 0.3. Therefore, the amplitudes of the entropic force and potential in water are significantly larger.

When the large sphere is replaced by a polyatomic solute, the structure of the solute also changes as the solute approaches the wall so that the entropic potential can be minimized. This effect should be significant especially when the entropic potential is positive and large. The free-energy barrier for the solute in contact with the wall to overcome by the thermal fluctuation can be considerably lower than in the case of the large sphere [21].

2.6 Roles of Electrostatic Interaction in Aqueous Solution Under Physiological Condition

In a microscopic self-assembly process in a biological system, when a portion with a positive net charge (portion A) in a biomolecule comes in contact with a portion with a negative net charge (portion B) in the same biomolecule or another biomolecule, energetic stabilization occurs due to electrostatic attractive interaction between portions A and B. However, this stabilization is accompanied by the loss of electrostatic attractive interactions between portion A and oxygen atoms carrying negative partial charges in water molecules and between portion B and hydrogen atoms carrying positive partial charges in water molecules, causing energetic destabilization. This destabilization is almost halved because the structure of some of the water near portions A and B is reorganized upon the contact (i.e., some of the water-water hydrogen bonds are recovered) [22]. Nevertheless, the net energy change relevant to water is positive and referred to as the "energetic dehydration penalty". This penalty, which is very large, is almost cancelled out by the stabilization energy described above. The contact of oppositely charged portions is quite important for compensating the energetic dehydration penalty. The contact of portions with positive net charges, for instance, causes energetic destabilization arising from not only electrostatic repulsive interaction between these portions but also the energetic dehydration penalty caused by the loss of electrostatic attractive interactions between these portions and oxygen atoms in water molecules. It is now decisive that the contact of portions possessing net charges in the same sign can hardly occur.

A receptor-ligand binding (binging of two solute molecules) is a good example of biological self-assembly processes. It is illustrated in Fig. 2.8 [23]. The binding is accompanied by a decrease in conformational energy of solute molecules and the energetic dehydration penalty explained above (see Fig. 2.8a). The decrease and the penalty are compensating (using our accurate statistical-mechanical theory [22]

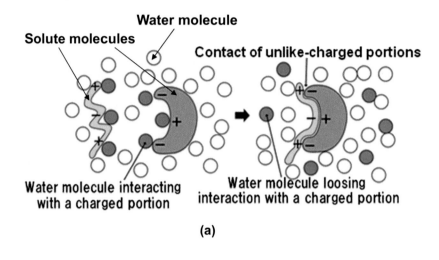

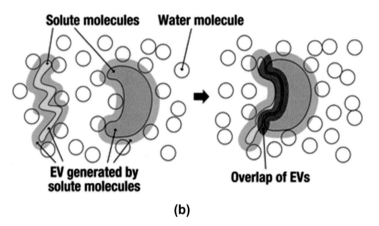

Fig. 2.8 Energetic (**a**) and entropic (**b**) aspects of binding of two solute molecules in water. In (**a**), the water molecules marked in blue are released to the bulk upon the binding with the result that some of the water-water hydrogen bonds are recovered. In (**b**), "EV" denotes the excluded volume. (Reproduced and modified from Ref. 24 with permission from the PCCP Owner Societies)

outlined in Chap. 6, we showed that the penalty is often larger than the decrease as exemplified in a protein-peptide binding process [24]).

The energetic dehydration penalty comprises its electrostatic (ES) and van der Waals (vdW) components. The penalty explained above is the ES component. The conformational energy of solute molecules also comprises its ES and vdW components. The binding is accompanied by not only a decrease in the vdW component of conformational energy but also the vdW component of the penalty arising from the loss of solute-water vdW interaction. For the binding of an oncoprotein (MDM2) and the extreme N-terminal peptide region of a tumor suppressor protein p53 (p53NTD) [24], for example, the quantities being discussed are as follows ($T = 298$ K):

Decrease in ES component of conformational energy $\sim-514k_BT$,
ES component of energetic dehydration penalty $\sim537k_BT$,
Decrease in vdW component of conformational energy $\sim-108k_BT$,
vdW component of energetic dehydration penalty $\sim95k_BT$.

Though these quantities are quite large, the sum of the two ES components is only $\sim23k_BT$, and the sum of the two vdW components is only $\sim-13k_BT$. The sum of all the ES and vdW components is positive and as small as $\sim10k_BT$. These quantities were calculated using our state-of-the-art theoretical method where molecular models are adopted for water, the structures of biomolecules (the MDM2-p53NTD complex and isolated MDM2 and p53NTD) are treated at the atomic level, and the structural fluctuation of the biomolecules in water are taken into account with the aid of molecular dynamics (MD) simulations with all-atom potentials [22, 24] (see Chap. 6 for more details). Therefore, the calculated values are quantitatively reliable. We emphatically remark that by a dielectric continuum model of water, the entropic EV effect cannot be taken into consideration and the energetic dehydration penalty is not calculable with quantitative accuracy.

In aqueous solution under the physiological condition (water containing NaCl at a concentration of ~ 0.15 mol/L), the electrostatic interaction is screened by not only water molecules but also cations and anions. As a consequence, it becomes over two orders of magnitude weaker and much shorter-ranged than in vacuum [25]. The entropic interaction mentioned above, which originates from the translational displacement of water molecules, is significantly stronger than the screened electrostatic interaction.

In a biological self-assembly process, charged portions in a biomolecule or biomolecules are driven to become buried by the water-entropy effect. For the contact of oppositely charged portions to occur upon the burial, the barrier due to the energetic dehydration penalty mentioned above must be overcome. It is overcome by a large gain of water entropy. The contact of oppositely charged portions is quite important, but it does not work as a significant driving force in a microscopic self-assembly process. Unfortunately, this is not well recognized in the biophysical and biochemical research communities.

2.7 Translational, Configurational Entropy of Water Leading Receptor-Ligand Binding and Protein Folding

A receptor-ligand binding (binging of two solute molecules), a good example of biological self-assembly processes, is accompanied by a water-entropy gain and a loss of conformational entropy of solute molecules (see Fig. 2.8b). However, the gain is significantly larger than the loss [23, 24, 26]. For the binding of MDM2 and p53NTD [24], the water-entropy gain is ~$132k_B$ and the conformational-entropy loss is $-59k_B$. The water-entropy gain was calculated using our recently developed, accurate method where molecular models are adopted for water and the structures of biomolecules are treated at the atomic level [22]. The conformational-entropy loss was calculated by means of the Boltzmann-quasi-harmonic method [27] combined with MD simulations with all-atom potentials. The calculated values are quantitatively reliable. The gain is more than twice larger than the loss. We can conclude that the water-entropy gain is a dominant contributor to the binding free energy. The number of residues of p53NTD is 12. When p53NTD is replaced by another peptide with 12 residues (actually, two peptides were tested in our earlier work [24]), the conformational-entropy loss remains almost unchanged but the water-entropy gain changes to a significant extent. Thus, the water-entropy gain is much more sensitive to the peptide properties and important as the key quantity governing the binding affinity of a peptide for the protein.

Protein folding is another good example of biological self-assembly processes. Protein folding unavoidably undergoes an energetically unfavorable loss of protein-water hydrogen bonds, but the formation of intramolecular hydrogen bonds using α-helix and β-sheet works as a factor opposing the loss. Moreover, as illustrated in Fig. 2.9, the formation of α-helix by a portion of the backbone or the formation of β-sheet by a lateral contact of portions of the backbone leads to a reduction of the EV followed by a water-entropy gain. Thus, the formation of α-helix and β-sheet is favorable both energetically and entropically [28–31]. This is why the folded state of a protein possesses significantly large percentages of α-helix and β-sheet. On the other hand, close packing of side chains with diverse geometric characteristics is crucially important. The presence of a side chain generates an excluded space which is inaccessible to the centers of water molecules. When the side chains are closely packed as illustrated in Fig. 2.9, the excluded spaces overlap with the result of a large gain of water entropy [28–31]. A protein folds so that the backbone and side chains (i.e., the protein atoms) can closely be packed with the formation of as much α-helix and β-sheet as possible (see Fig. 2.9).

We demonstrated that the folding of apoplastocyanin (apoPC) with 99 residues at 298 K is characterized by a very large water-entropy gain of ~$670k_B$ [32]. This gain surpasses the conformational-entropy loss of the protein (~$-300k_B$) and a positive enthalpy change upon the folding, and the decrease in system free energy upon the folding is ~$-20k_BT$ ($T = 298$ K) [32]. The water-entropy gain is over twice larger than the conformational-entropy loss. The gain of protein intramolecular hydrogen bonds is unavoidably accompanied by the loss of protein-water hydrogen bonds and the recovery of some of water-water hydrogen bonds, giving rise to the energetic dehydration penalty [23–26]. The positive enthalpy change, which was measured at 298 K using a novel experimental technique [33], is ascribed to this penalty. An important conclusion is that the water-entropy gain drives a protein to fold. The water-entropy gain consists of the gains of translational, configurational entropy and rotational entropy. However, we showed that the gain of translational, configurational entropy is much larger [32].

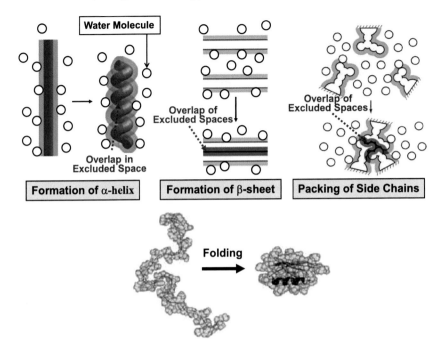

Fig. 2.9 Top: Formation of α-helix (left) and β-sheet (middle) in backbone and close packing of side chains (right) in protein folding. It should be noted that α-helix and β-sheet are advantageous structural units in terms of the water entropy as well as the intramolecular hydrogen bonding. The close packing of side chains leads to a large gain of water entropy. Bottom: Protein folding. The backbone and side chains are closely packed with the formation of as much α-helix and β-sheet as possible

In many studies, only compact structures are considered for a protein. For a structural transition from a fully extended structure to a compact one, the water-entropy gain is much more sensitive to characteristics of the compact structure than the conformational-entropy loss. Therefore, if unrealistic compact structures with the high energetic dehydration penalty are excluded, the native structure can be discriminated from a number of nonnative, compact structures using only the water-entropy gain as a criterion function [30, 31].

For some proteins, the amino-acid sequence is optimized so that the backbone and side chains can closely be packed like a three-dimensional jigsaw puzzle. In other proteins, however, this type of overall close packing is not achievable. In such cases, only portions which are amenable to close packing followed by a significantly large water-entropy gain are preferentially packed. In yeast frataxin [34] shown in Fig. 2.10, for instance, the preferential, close packing is accomplished by excluding some portions as a tail and by forming a vacant space. Impartial, less close packing causes a larger EV generated by the protein, which is less favorable in terms of the water entropy. For a protein complex, the most important requirement is to achieve closely packed interfaces between proteins in the complex.

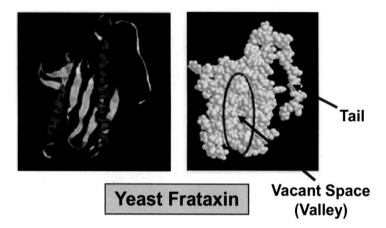

Fig. 2.10 Structure of yeast frataxin [33]. Left: Ribbon representation. Right: Space-filled representation. The backbone and side chains are closely packed by excluding the tail, and the formation of the vacant space cannot be avoided in the close packing

2.8   Essential Roles of Water-Entropy Effect in Biological Processes

As argued above, the gain of water entropy drives a variety of biological self-assembly processes such as protein folding [28–32], different types of molecular recognition [23, 24, 26, 35], and formation of protein aggregates such as amyloid fibrils [13]. Moreover, the dependence of water-entropy effect on T and P is much stronger than that of the other enthalpic components [36]. The water-entropy effect provides a clue to the mechanism of pressure [37, 38] and cold [37, 39] denaturing of a protein. Importantly, the presence of a water molecule also generates an EV for the other water molecules, all the water molecules in the system are thus entropically correlated. We refer to this entropic correlation as the "water crowding". The water crowding becomes more significant as the total EV generated by the biomolecules immersed in water increases. Upon a self-assembly process, the total EV is reduced and the water crowding is mitigated, leading to a water-entropy gain. This is the true physical origin of the hydrophobic effect [14, 15].

Here are intriguing examples for which the entropic force or potential generated by water is a physical factor of paramount importance. Insertion of a solute into a vessel consisting of biopolymers followed by release of the solute from the vessel is a fundamental function in a biological system. We introduce the following two paradigmatic examples: (I) After an unfolded protein is inserted into chaperonin GroEL from bulk aqueous solution, protein folding is finished within the cavity of GroEL, and the folded protein is released back to the bulk aqueous solution [14, 40]; and (II) after an antibiotic molecule (a substrate) is inserted into ABC transporter from the inside of cell membrane, it is released from the transporter to the outside [14, 41]. The switch from insertion to release is achieved by the change in the structure and properties of the protein upon the folding in (I) and by the structural change of the transporter in (II) (see Fig. 1.1c). Diverse proteins are inserted and released in (I), and a variety of substrates are carried across the membrane in (II). We showed that the entropic force and potential generated by water play essential roles in the insertion process in (I) [14, 40] and the insertion and release processes in (II) [14, 41–43]. (An energetic factor plays a pivotal role in the release process in (I) [14, 40].)

A molecular motor, an ATP-driven protein or protein complex, functions in aqueous solution under the physiological condition. As described in Sect. 2.5.1, the entropic EV effect is remarkably large for a protein or protein complex immersed in water. The importance of the translational, configurational entropy of water over the electrostatic interaction, which is argued above, holds true for the functional expression of the molecular motor as well.

2.9 Recent Papers Pointing Out Crucial Importance of Hydration Effect on Unidirectional Movement of Myosin Along F-Actin

In the literature, there are two pioneering papers [44, 45] pointing out the essential roles of hydration effect in the unidirectional movement of myosin along F-actin. The first [44] and second [45] papers were written by us and Suzuki et al., respectively. S1 was considered as a simple but physically insightful model of myosin. The claim shared by these papers is that the force for moving myosin is generated by water.

First, we review the second paper [45] (it was revisited in a recently published book [46]). The key quantity is claimed to be the hydration free energy of S1, μ, which takes a significantly large, negative value. After ATP bound to S1 is hydrolyzed into ADP and Pi, S1 weakly binds to F-actin. During this weak binding, S1 repeats the detachment from and the attachment to F-actin. While S1 is detached from F-actin, S1 is driven to move in the right direction for the following reason. S1 feels the negative electric field emanated from F-actin. Importantly, μ becomes lower (i.e., $|\mu|$ becomes larger) as the field strength increases. The structure of F-actin on the left side of S1 is different from that on the right side of S1, and the electric field emanated from F-actin on the right side of S1 is stronger than that on the left side of S1. Therefore, S1 moves in the right direction so that μ can be lowered.

We then comment on the second paper [45, 46]. The quantity to be looked at is not the hydration free energy of S1 but the system free energy represented by the sum of the hydration free energy (a), conformational energy (b) (i.e., energy in vacuum), and conformational entropy (c) of the protein complex (i.e., actomyosin) comprising S1 and F-actin. S1 is moved so that the system free energy can be lowered. The hydration free energy can be decomposed into the hydration energy (a−1) and entropy (a−2). We note that factors (a−1) and (b) are compensating: When (a−1) becomes lower, for example, (b) always becomes higher. Factor (c) is relatively smaller. Therefore, the change in system free energy can be approximated by that in factor (a−2). It turns out that the key quantity is the hydration entropy of the protein complex as emphasized in the first paper [44] and revisited in a recently published book [47]. Our claim is that the entropic force generated by water is a dominant physical factor, and the geometric characteristics of F-actin are more important than the electric-field distribution emanated from F-actin.

2.10 Problems in Prevailing View on Functional Expression of a Molecular Motor

No one considers that a protein must perform mechanical work against the viscous resistance force by water during the folding. Water is not the external system: The system of interest consists of not only the protein but also water where it is immersed. The protein folds so that the free energy of the protein-water system can be minimized. A self-assembly process such as protein folding spontaneously occurs and no input of energy or free energy is necessitated. Water never hinders protein folding through the viscous resistance force: It does drive a protein to fold as argued in Sect. 2.7.

Here, we comment on the physical difference between the EV, V_{ex}, and the partial molar volume (PMV), V_M [38]. V_M of a solute is the change in system volume upon insertion of the solute into water under the isobaric condition. V_{ex} is determined only by the geometric characteristics of protein structure and the molecular diameter of water, but V_M is also dependent on the average number density of water molecules near the protein surface and the water-accessible-surface area (WASA) [38]. A higher-density layer of water molecules is formed near the protein surface. The number density of water molecules within this layer is significantly higher than that in bulk water. As the WASA increases, "$V_{ex}-V_M$", which is positive, becomes larger. V_{ex} of the folded state is much smaller than that of the unfolded state. However, the WASA of the folded state is much smaller than that of the unfolded state. Consequently, the values of V_M of the folded and unfolded states are not significantly different (in general, V_M of the folded state is slightly larger than that of the unfolded state) [32, 33]. (A pressure-denatured state is unique in the sense that its EV is only slightly larger than the EV of the native state while its WASA is considerably larger than the WASA of the native state. As a result, the PMV of the pressure-denatured state is significantly smaller than that of the native state [38].) Hence, the system volume does not change much during the folding under the isobaric condition. Moreover, the pressure P is only 1 atm. It follows that $P\Delta V \sim 0$ (ΔV is the change in system volume and $P\Delta V$ is the mechanical work performed by the system) and the system performs essentially no mechanical work. This is true for any biological self-assembly process. For the binding of MDM2 and p53NTD [24], $0 < P\Delta V \ll k_B T$ ($T = 298$ K). For the apoPC folding, $0 < P\Delta V \ll k_B T$ is corroborated both experimentally [33] and theoretically [32].

For actomyosin, for instance, the system of interest consists of not only myosin and F-actin but also water in which ATP, ADP, and Pi are dissolved. Hydration of myosin and F-actin plays essential roles in their behavior. During the movement of myosin along F-actin, the PMVs of myosin and F-actin do not change much and $P\Delta V \sim 0$ ($P = 1$ atm), with the result that the system performs essentially no mechanical work. The concept, "myosin must perform mechanical work against the viscous resistance force by water during the movement", is incorrect. It is water that drives myosin to move along F-actin [14, 44, 47]. The unidirectional movement of myosin along F-actin, which is coupled with the ATP hydrolysis reaction (an irreversible process accompanied by a decrease in system free energy), spontaneously occurs. Taken together, the prevailing view, which distinguishes the functional expression of a molecular motor from self-assembly processes such as protein folding, is problematic and must be reconsidered.

2.11   Inconsistency of Prevailing View with Some of Recent Experimental Facts

The prevailing view on the functional expression of a molecular motor is problematic as theoretically pointed out in Sect. 2.10. What is worse, it does not coincide with some of recent experimental observations. Iwaki et al. [48] experimentally studied the unidirectional movement of myosin along F-actin in aqueous solution to which sucrose was added. First, a sucrose concentration of 1 mol/L was tested. By this sucrose addition, the viscosity of aqueous solution became about six times higher and the viscous resistance force by water became about six times stronger. If the prevailing view was correct, the movement of myosin would be stopped or remarkably affected. The result observed was that the sucrose addition has virtually no effects on the myosin movement. When the sucrose concentration was increased to 2 mol/L, the myosin movement was stopped. However, this stop was shown to be attributable to significantly stronger binding of myosin to F-actin [48]. In water, myosin gets detached from F-actin upon the ATP binding to myosin. In aqueous solution of sucrose at 2 mol/L, on the other hand, the ATP binding does not lead to the detachment with the result that the myosin movement is stopped.

Addition of a highly hydrophilic cosolvent such as sucrose increases the packing fraction of aqueous solution. Therefore, the addition of sucrose enlarges the water-entropy effect: It enhances the thermostability of the folded state of a protein [49] but never hinders protein folding. In the case of actomyosin, the entropic potential field acting on myosin along the y-axis (see Fig. 2.11) due to the presence of F-actin, $\Phi_{\text{F-actin}}(y)$, is qualitatively similar to $\Phi_{\text{Wall}}(h)$ shown in Fig. 2.6b. In water containing no sucrose, myosin without ATP bound cannot overcome the free-energy barrier by the thermal fluctuation for getting detached from F-actin [47]. Upon the ATP binding to myosin, the structure of myosin and $\Phi_{\text{F-actin}}(y)$ change with the result of the reduction in the barrier, and myosin can readily get detached from F-actin [47]. However, the amplitudes of $\Phi_{\text{F-actin}}(y)$ and the barrier are made larger by the sucrose

Fig. 2.11 Choice of y- and x-axes. The entropic potential field acting on myosin along the y-axis due to the presence of F-actin, $\Phi_{\text{F-actin}}(y)$, is qualitatively similar to $\Phi_{\text{Wall}}(h)$ shown in Fig. 2.6b

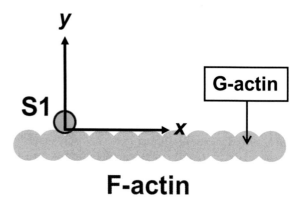

addition, and even myosin with ATP bound can hardly overcome the barrier. Taken together, the experimental result of Iwaki et al. mentioned above does not coincide with the prevailing view.

The above discussion emphasizing the water-entropy effect is based on the experimental data [2, 50] manifesting the following: (i) The binding entropy and enthalpy between myosin and F-actin are both positive and therefore the binding is entropically driven (the binding entropy is the entropy gain occurring upon myosin binding to F-actin); (ii) the binding entropy for myosin with ATP bound and F-actin is much smaller than that for myosin without ATP bound and F-actin; and (iii) the binding affinity of myosin for F-actin is substantially weakened at low temperatures [50]. Result (ii) is consistent with the experimentally known behavior that myosin can get detached from F-actin only after ATP binds to it. Result (iii), which originates from the weakening of the water-entropy effect (i.e., the weakening of the hydrophobic effect) [14, 15, 44], is associated with the cold denaturation of a protein [37, 39]. At low temperatures, the hydrogen bonding of water molecules is strengthened, with the result that the translational motion of water molecules becomes less active and water crowding in the bulk is mitigated. For this reason, the entropic power of driving a self-assembly process becomes significantly weaker. (A review of the hydrophobic effect is provided in our recent article [51]).

Experimental studies using high-speed atomic force microscopy [52, 53] showed with surprise that an input of ATP energy or free energy is not required for the unidirectional movement of myosin. Very recently, Kodera and coworkers [53] have obtained the following striking result: Even in the absence of ATP and ATP hydrolysis cycle, when the head of the hind leg (trailing head) of myosin V is artificially detached from F-actin using an advanced technique, myosin V makes a unidirectional, forward walk. This result is in accord with our view that the force which makes myosin move unidirectionally is generated by not ATP but water. The conclusion drawn by Kodera and coworkers was that the input of ATP energy or free energy was required for the detachment of myosin from F-actin [52, 53]. However, this conclusion is not correct. First, such an input is not necessitated for any elementary process in the unidirectional movement. Second, the detachment stems from the structural change of myosin upon the ATP binding followed by the aforementioned change in $\Phi_{\text{F-actin}}(y)$.

References

1. Kodama T (2018) Energetics of Myosin ATP hydrolysis by calorimetry. In: Suzuki M (ed) The role of water in ATP hydrolysis energy transduction by protein machinery, Chapter 7, Part II, Springer Briefs in Molecular Science, Springer, ISBN: 978-981-10-8458-4, pp 103–111 (2018)
2. Kodama T (1985) Physiol Rev 65:467
3. Alberty RA (1994) Pure Appl Chem 66:1641
4. Voet D, Voet JG (2004) Biochemistry, 3rd edn. Wiley, New York
5. Kitamura K, Tokunaga M, Iwane AH, Yanagida T (1999) Nature 397:129
6. Kitamura K, Tokunaga M, Esaki S, Iwane AH, Yanagida T (2005) Biophysics 1:1
7. Ward A, Reyes CL, Yu J, Roth CB, Chang G (2007) Proc Natl Acad Sci USA 104:19005
8. Hollenstein K, Dawson RJP, Locher KP (2007) Curr Opin Struct Biol 17:412
9. Shimabukuro K, Yasuda R, Muneyuki E, Hara KY, Kinosita K Jr, Yoshida M (2003) Proc Natl Acad Sci USA 100:14731
10. Adachi K, Oiwa K, Nishizaka T, Furuike S, Noji H, Itoh H, Yoshida M, Kinosita K Jr (2007) Cell 130:309
11. Kinoshita M (2002) J Chem Phys 116:3493
12. Kinoshita M (2006) Chem Eng Sci 61:2150
13. Kinoshita M (2004) Chem Phys Lett 387:54
14. Kinoshita M (2016) Mechanism of functional expression of the molecular machines. Springer Briefs in Molecular Science, Springer, ISBN: 978-981-10-1484-0
15. Kinoshita M (2013) Biophys Rev 5:283
16. Soda K (1993) J Phys Soc Jpn 62:1782
17. Asakura S, Oosawa F (1954) J Chem Phys 22:1255
18. Asakura S, Oosawa F (1958) J Polym Sci 33:183
19. Hansen J-P, McDonald LR (2006) Theory of simple liquids, 3rd edn. Academic Press, London
20. Kinoshita M, Hayashi T (2017) Phys Chem Chem Phys 19:25891
21. Kinoshita M (unpublished results)
22. Hikiri S, Hayashi T, Inoue M, Ekimoto T, Ikeguchi M, Kinoshita M (2019) J Chem Phys 150:175101
23. Hayashi T, Matsuda T, Nagata T, Katahira M, Kinoshita M (2018) Phys Chem Chem Phys 20:9167
24. Yamada T, Hayashi T, Hikiri S, Kobayashi N, Yanagida H, Ikeguchi M, Katahira M, Nagata T, Kinoshita M (2019) J Chem Inf Model 59:3533
25. Kinoshita M, Harano Y (2005) Bull Chem Soc Jpn 78:1431
26. Hayashi T, Oshima H, Mashima T, Nagata T, Katahira M, Kinoshita M (2014) Nucleic Acids Res 42:6861
27. Hikiri S, Yoshidome T, Ikeguchi M (2016) J Chem Theory Comput 12:5990
28. Yasuda S, Yoshidome T, Oshima H, Kodama R, Harano Y, Kinoshita M (2010) J. Chem. Phys. 132:065105
29. Yasuda S, Oshima H, Kinoshita M (2012) J Chem Phys 137:135103
30. Hayashi T, Yasuda S, Škrbić T, Giacometti A, Kinoshita M (2017) J Chem Phys 147:125102
31. Hayashi T, Inoue M, Yasuda S, Petretto E, Škrbić T, Giacometti A, Kinoshita M (2018) J Chem Phys 149:045105

32. Yoshidome T, Kinoshita M, Hirota S, Baden N, Terazima M (2008) J Chem Phys 128:225104
33. Baden N, Hirota S, Takabe T, Funasaki N, Terazima M (2007) J Chem Phys 127:175103
34. Amano K, Yoshidome T, Harano Y, Oda K, Kinoshita M (2009) Chem Phys Lett 474:190
35. Hayashi T, Oshima H, Yasuda S, Kinoshita M (2015) J Chem Phys B 119:14120
36. Kinoshita M, Oshima H (2014) Chem Phys Lett 610–611:1
37. Oshima H, Kinoshita M (2015) J Chem Phys 142:145103
38. Inoue M, Hayashi T, Hikiri S, Ikeguchi M, Kinoshita M (2020) J Chem Phys 152:065103
39. Inoue M, Hayashi T, Hikiri S, Ikeguchi M, Kinoshita M (2020) J Mol Liq 317:114129
40. Amano K, Oshima H, Kinoshita M (2011) J Chem Phys 135:185101
41. Mishima H, Oshima H, Yasuda S, Amano K, Kinoshita M (2013) J Chem Phys 139:205102
42. Amono K, Kinoshita M (2010) Chem Phys Lett 488:1
43. Mishima H, Oshima H, Yasuda S, Amano K, Kinoshita M (2013) Chem Phys Lett 561–562:159
44. Amano K, Yoshidome T, Iwaki M, Suzuki M, Kinoshita M (2010) J Chem Phys 133:045103
45. Suzuki M, Mogami G, Hideyuki Ohsugi, Watanabe T, Matubayasi N (2017) Cytoskeleton 74:512
46. Suzuki M, Mogami G, Watanabe T, Matubayasi N (2018) Novel intermolecular surface force unveils the driving force of the actomyosin system. In: Suzuki M (ed) The role of water in ATP hydrolysis energy transduction by protein machinery. Chapter 16, Part III, Springer Briefs in Molecular Science, Springer, ISBN: 978-981-10-8458-4, pp 257–274
47. Kinoshita M (2018) Functioning mechanism of ATP-driven proteins inferred on the basis of water-entropy effect. In: Suzuki M (ed) The role of water in ATP hydrolysis energy transduction by protein machinery. Chapter 18, Part III, Springer Briefs in Molecular Science, Springer, ISBN: 978-981-10-8458-4, pp. 303-323
48. Iwaki M, Ito K, Fujita K (2018) Single-molecule analysis of actomyosin in the presence of osmolyte. In: Suzuki M (ed) The Role of Water in ATP Hydrolysis Energy Transduction by Protein Machinery. Chapter 15, Part III, Springer Briefs in Molecular Science, Springer, ISBN: 978-981-10-8458-4, pp. 245-256
49. Oshima H, Kinoshita M (2013) J Chem Phys 138:245101
50. Katoh T, Morita F (1996) J Biochem 120:189
51. Yasuda S, Kazama K, Akiyama T, Kinoshita M, Murata T (2020) J Mol Liq 301:112403
52. Kodera N, Ando T (2014) Biophys. Rev. 6:237
53. Kodera N et al. (unpublished results)

Chapter 3
Mechanism of Unidirectional Rotation of γ Subunit in F_1-ATPase

Abstract In this chapter, the discussion is focused on the functional expression of F_1-ATPase, the unidirectional rotation of the γ subunit. In the structure of the $\alpha_3\beta_3$ complex stabilized by the water-entropy effect, the atoms in three β subunits, to which different chemical compounds (i.e., ATP just before the hydrolysis reaction, ATP, ADP + Pi, nothing, or Pi) are bound, are closely, moderately, and loosely packed, respectively. This nonuniformity and the structural asymmetry of the γ subunit play essential roles. The γ subunit takes a particular orientation in accordance with the structure of the $\alpha_3\beta_3$ complex so that the water entropy can be maximized. Due to the occurrence of the ATP binding, ATP hydrolysis, and dissociation of ADP and Pi during each ATP hydrolysis cycle, the chemical compounds bound to the three β subunits successively change, in concert with which the $\alpha_3\beta_3$ complex structure and the orientation of the γ subunit sequentially change to retain the maximized water entropy. In one hydrolysis cycle, the γ subunit exhibits a 120° rotation. It is experimentally known that the ATP synthesis occurs when the γ subunit is forced to rotate in the inverse direction by a sufficiently strong external torque imposed. This can also be explicated on the basis of the water-entropy effect.

Keywords F_1-ATPase · Nucleotide · ATP hydrolysis · Entropic force · Entropic potential · Unidirectional rotation · ATP synthesis

3.1 Definition of Packing Structure for a Protein or Protein Complex

The free energy for a biomolecule immersed in water is represented by the sum of conformational energy, hydration energy, conformational entropy, and hydration entropy. As argued in Sects. 2.6–2.8, upon a structural change of a protein or protein complex or a binding of two biomolecules, the change in hydration entropy is much larger than the sum of the changes in conformational energy and in hydration energy and significantly larger than the change in conformational entropy. Therefore, the change in hydration entropy (or equivalently, the change in water entropy) is a principal component of the free-energy change. In Chap. 3, we elucidate the mechanism of unidirectional rotation of the γ subunit in F_1-ATPase with emphasis on the water-entropy effect. It is worthwhile to recall that unlike in vacuum the electrostatic interaction is much less important than one might expect (see Sects. 2.6 and 3.4.5).

Hereafter, we often use the term "packing structure" for a protein or protein complex. As the packing of protein atoms (i.e., atoms in the backbone and side chains) becomes closer, the EV generated by the protein reduces, which is more favorable in terms of the water entropy. We also use the term "packing efficiency" signifying the degree of close packing. Closer packing can be referred to as "higher packing efficiency" (when the atoms are more closely packed, we state that they are packed with higher efficiency). For a protein complex, not only the atoms in each protein in the complex but also those in the interface between each protein pair needs to be more closely packed. In general, however, it is not possible to meet both of these requirements by uniform packing, i.e., by impartial packing of all the atoms constituting the complex. As discussed for a protein in Sect. 2.7, it is often that nonuniform packing is more favorable than the uniform packing. Here, the nonuniform packing implies that only the atoms amenable to close packing are preferentially packed with higher efficiency and the packing of the other atoms is left looser. For a protein complex, we refer to the differences in the packing efficiency among the proteins forming the complex and among the protein interfaces as the "packing structure".

3.2   Nonuniform Binding of Nucleotides to $\alpha_3\beta_3$ or $\alpha_3\beta_3\gamma$ Complex

We first review the experimentally determined structures of the $\alpha_3\beta_3$ complex. In the absence of nucleotides in aqueous solution, the $\alpha_3\beta_3$ complex takes a structure with three-fold symmetry as shown in Fig. 3.1a [1]. The structures of the three β subunits are the same: All of them take open structure [1, 2]. In aqueous solution of AMP-PNP, an analogue of ATP whose hydrolysis reaction does not occur, AMP-PNP is bound to two of the three β subunits and nothing is bound to the other β subunit as shown in Fig. 3.1b [1]. Each of the β subunits to which AMP-PNP is bound takes closed structure whereas the β subunit to which nothing is bound takes open structure. The packing of atoms constituting the β subunit in the closed structure should be closer than that in the open structure.

We then review the experimentally determined structures of the $\alpha_3\beta_3\gamma$ complex. According to the crystal structures of F_1-ATPase from bovine heart mitochondria in aqueous solution where AMP-PNP and ADP are dissolved as nucleotides, AMP-PNP is bound to two of the three β subunits and nothing is bound to the other β subunit in the absence of azide (see Fig. 3.2a), and AMP-PNP, ADP, and nothing are bound to the three β subunits, respectively, in the presence of azide [3] (see

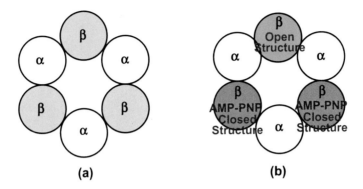

Fig. 3.1 **a** Structure of $\alpha_3\beta_3$ complex stabilized in aqueous solution containing no nucleotides. **b** Structure of $\alpha_3\beta_3$ complex stabilized in aqueous solution of AMP-PNP. The packing of atoms constituting the β subunit in the closed structure is closer than that in the open structure

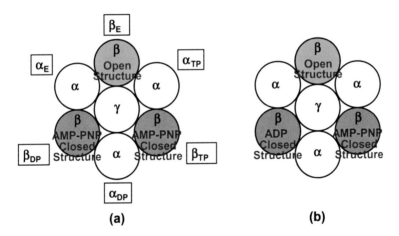

Fig. 3.2 **a** Structure of $\alpha_3\beta_3\gamma$ complex stabilized in aqueous solution of AMP-PNP and ADP. **b** Structure of $\alpha_3\beta_3\gamma$ complex stabilized in aqueous solution of AMP-PNP, ADP, and azide. Azide is known to stabilize the β subunit to which ADP is bound [3]. The three β subunits are named β_{DP}, β_{TP}, and β_E, respectively, as shown in (**a**). β_{DP}: The β subunit to which AMP-PNP is bound in (**a**) and that to which ADP is bound in (**b**). β_{TP}: The β subunit to which AMP-PNP is bound in (**a**) and (**b**). β_E: The β subunit to which nothing is bound in (**a**) and (**b**). The three α subunits are named α_E, α_{TP}, and α_{DP}, respectively, as shown in (**a**)

Fig. 3.2b). A more detailed picture of the structure shown in Fig. 3.2a is presented in Fig. 3.3. The three β subunits are named β_{DP}, β_{TP}, and β_E, respectively, and the three α subunits are named α_E, α_{TP}, and α_{DP}, respectively, as shown in Fig. 3.2a. β_{DP} and β_{TP} take closed structure whereas β_E takes open structure. The structure reported by Abrahams et al. [4] also looks like that shown in Fig. 3.2b. In the structure of the nucleotide-free $\alpha_3\beta_3\gamma$ complex, one of the three β subunits takes open structure and the structures of the other two β subunits are closed (see Fig. 3.4) [5]. This suggests

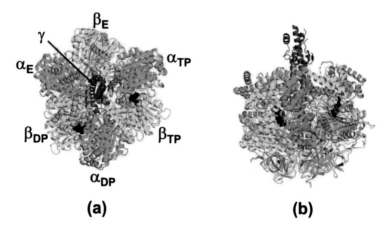

Fig. 3.3 Ribbon representation of α$_3$β$_3$γ-complex structure shown in Fig. 3.2a. **a** Top view. **b** Side view. The α subunits, β subunits, and γ subunit are colored green, yellow, and gray, respectively. AMP-PNP is represented by the red fused spheres

Fig. 3.4 Structure of α$_3$β$_3$γ complex to which nucleotides are not bound [5]. The α$_3$β$_3$γ complex is immersed in aqueous solution containing no nucleotides

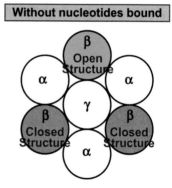

that the interaction between the γ subunit and the α$_3$β$_3$ complex is as important as the nucleotide occupancy in determining the structural state of the three β subunits.

Taken together, the α$_3$β$_3$ or α$_3$β$_3$γ complex stabilized (i.e., irrespective of the presence of the γ subunit) is characterized by two β subunits possessing closed structure with nucleotides bound and one β subunit possessing open structure without nucleotides bound. Presumably, if all the three β subunits took closed structure, it would become impossible to form sufficiently close overall packing of the atoms in the six α–β interfaces, which was less favorable in terms of the water entropy. It is important to note that the incorporation of the γ subunit in the α$_3$β$_3$ complex also leads to the state characterized by two β subunits possessing closed structure and one β subunit possessing open structure, definitely due to the structural asymmetry inherent in the γ subunit. The specific packing structure of the α$_3$β$_3$ complex is stabilized either by the nonuniform nucleotide binding or the incorporation of the γ subunit, but it is further stabilized when both of them are conferred upon the complex.

In aqueous solution of ATP, ADP, and Pi, the structural rotation in the counterclockwise direction (in this book, the direction of the rotation is discussed by viewing F_1-ATPase from the F_o side) illustrated in Fig. 3.5 occurs in the $\alpha_3\beta_3$ complex even without the rotor, the γ subunit [1]. This experimental result indicates that the rotation mechanism is programmed even in the $\alpha_3\beta_3$ complex itself. It is probable that each of the states illustrated in Fig. 3.5a is the catalytic dwell state [3, 6–8] illustrated in Fig. 3.6, where ATP, ATP just before the hydrolysis reaction, and nothing are bound to the three β subunits, respectively. In Fig. 3.6, "ATP• • •H_2O" denotes ATP just before the hydrolysis reaction (the activated complex). As argued in Sect. 3.3, the catalytic dwell state is highly stable in terms of the water entropy. In other words, the binding process illustrated in Fig. 3.7a should lead to the largest gain of water entropy. The water-entropy gain upon the process in Fig. 3.7a is larger than the other binding processes such as the process illustrated in Fig. 3.7b. This result should be true even for the $\alpha_3\beta_3$ complex.

In summary, the water-entropy effect stabilizes a specific packing structure of the $\alpha_3\beta_3\gamma$ complex in aqueous solution of ATP, ADP, and Pi through not only the

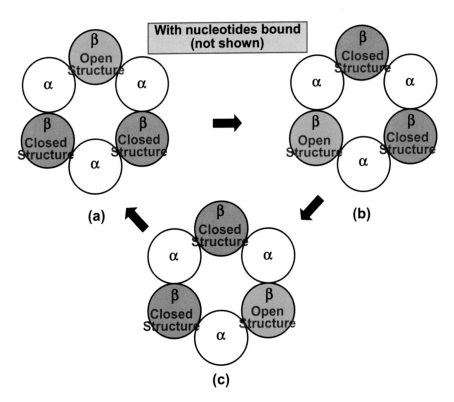

Fig. 3.5 Structural rotation of $\alpha_3\beta_3$ complex to which nucleotides are bound [1]. The $\alpha_3\beta_3$ complex rotates in the counterclockwise direction (a)→(b)→(c)→(a) in aqueous solution of ATP, ADP, and Pi. The solution is under the condition that the ATP hydrolysis reaction occurs

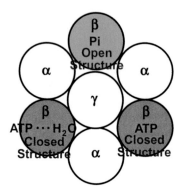

Fig. 3.6 Structure of catalytic dwell state of $\alpha_3\beta_3\gamma$ complex stabilized in aqueous solution of ATP, ADP, and Pi. "ATP•••H_2O" represents ATP just before the hydrolysis reaction (i.e., the activated complex)

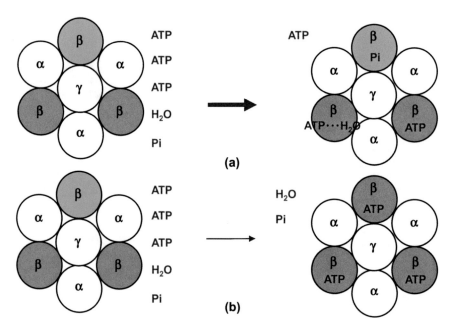

Fig. 3.7 a Binding of ATP, ATP, and Pi to three β subunits in aqueous solution of ATP, ADP, and Pi. ATP is bound to a β subunit as ATP•••H_2O (see the caption for Fig. 3.6). **b** Binding of ATP, ATP, and ATP to the three β subunits in aqueous solution of ATP, ADP, and Pi

nonuniform binding of nucleotides to the three β subunits but also the structural asymmetry of the γ subunit. It is probable that the packing structure of the $\alpha_3\beta_3$ complex is determined primarily by the packing structures of the three β subunits. We emphatically remark that the packing efficiency in a β subunit is intimately related to the chemical compound (i.e., ATP•••H_2O, ATP, ADP + Pi, nothing, or Pi) bound

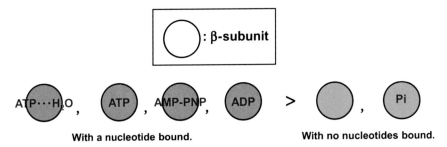

Fig. 3.8 Order of packing efficiency for β subunit to which ATP• • •H$_2$O (see the caption for Fig. 3.6), ATP, AMP-PNP, ADP, nothing, or Pi is bound. The packing efficiency of a β subunit to which ATP• • •H$_2$O, ATP, AMP-PNP, or ADP is bound (i.e., a β subunit with a nucleotide bound) is higher than that to which nothing or Pi is bound (i.e., a β subunit with no nucleotides bound)

to it. From the characteristics of the experimental structures reviewed above, we can infer that the packing efficiency of the β subunit should follow the order depicted in Fig. 3.8. The γ subunit possessing an asymmetric structure takes a specific orientation in response to the packing structure of the $\alpha_3\beta_3$ complex. We note that the structural stabilization is governed by the water-entropy effect. (The above discussion is made convincing by our theoretical analyses described in Sect. 3.3).

3.3 Theoretical Analyses on Packing Structure of $\alpha_3\beta_3\gamma$ Complex in Catalytic Dwell State

We consider the following four scenarios: (A) rotation in the normal direction under the solution condition that the ATP hydrolysis reaction occurs; (B) rotation in the inverse direction under the solution condition that the ATP synthesis reaction occurs; (C) rotations in random directions under the solution condition that the ATP hydrolysis and synthesis reactions are in equilibrium; and (D) occurrence of ATP synthesis, even under the solution condition that the ATP hydrolysis reaction should occur, through forcible rotation in the inverse direction by means of sufficiently strong external torque imposed on the central shaft. First, we analyze the packing structure of the catalytic dwell state [3, 6–8] in scenario (A) where [ATP] is sufficiently high and [ADP] and [Pi] are sufficiently low (see Sect. 2.1.1).

3.3.1 A State of $\alpha_3\beta_3\gamma$ Complex Stabilized: Catalytic Dwell State

The ATP hydrolysis reaction occurs under the solution condition assumed. Most of the experimentally solved structures of F$_1$-ATPase are for the catalytic dwell state

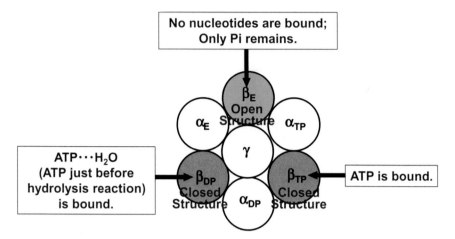

Fig. 3.9 Catalytic dwell state of $\alpha_3\beta_3\gamma$ complex. This is the initial state for one ATP hydrolysis cycle in scenario (A)

shown in Fig. 3.9 [3, 6–8] (see Fig. 3.6 also). It is quite stable in terms of the water entropy as discussed in Sect. 3.2. According to the experimental studies by Noji and coworkers [9], after the ATP hydrolysis to ADP and Pi in a β subunit, ADP first dissociates from the β subunit. In the catalytic dwell state, the remaining Pi, ATP, and ATP• •H₂O are bound to the β subunits denoted by β_E, β_{TP}, and β_{DP}, respectively. The three α subunits are named α_E, α_{TP}, and α_{DP}, respectively, as shown in Fig. 3.9. The $\alpha_3\beta_3\gamma$ complexes shown in Figs. 3.2a, b, and 3.6 or 3.9 share qualitatively the same characteristics of the packing structure.

3.3.2   Methods of Theoretical Analyses

We analyze the packing structure of the $\alpha_3\beta_3\gamma$ complex using the crystal structure of F₁-ATPase from bovine heart mitochondria (PDB ID: 2JDI) [3, 10] corresponding to the catalytic dwell state shown in Fig. 3.2a where Pi is not bound to β_E. The purpose of this analysis is to unveil the basic characteristics of the packing structure common in the $\alpha_3\beta_3\gamma$ complexes shown in Figs. 3.2a, b, and 3.6 or Fig. 3.9. The missing residues (402–409 in α_{TP}, 388–395 in β_E, 48–66, 87–104, 117–126, 149–158, and 174–205 in γ) are added using MODELLER [11] on the basis of the crystal structure whose PDB ID is 1E79 [12]. AMP-PNP and Mg²⁺ are bound to β_{TP} and β_{DP}, and no nucleotides are bound to β_E. For the theoretical analyses, AMP-PNP is replaced by ATP. Since there is no ATP-Mg²⁺ bound to β_E, β_E has significantly fewer atoms than the other two β subunits. To compare the packing efficiencies of proteins or protein interfaces impartially, we calculate the hydration entropy $S < 0$ of ATP-Mg²⁺ and add it to S of any protein, protein pair, or protein complex including β_E. Refer to our earlier publication [13] for more details.

Subcomplex I: β_E, α_E, α_{TP}, and γ.
Subcomplex II: β_{TP}, α_{TP}, α_{DP}, and γ.
Subcomplex III: β_{DP}, α_{DP}, α_E, and γ.

Subcomplex I$-\gamma$: β_E, α_E, and α_{TP}.
Subcomplex II$-\gamma$: β_{TP}, α_{TP}, and α_{DP}.
Subcomplex III$-\gamma$: β_{DP}, α_{DP}, and α_E.

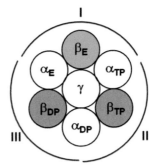

Fig. 3.10 Decomposition of $\alpha_3\beta_3\gamma$ complex into subcomplexes I, II, and III. Each subcomplex from which the γ subunit is removed is referred to as "subcomplex I$-\gamma$", "subcomplex II$-\gamma$", or "subcomplex III$-\gamma$"

The $\alpha_3\beta_3\gamma$ complex is decomposed into three subcomplexes as follows (see Fig. 3.10):

Subcomplex I: β_E, α_E, α_{TP}, and γ,
Subcomplex II: β_{TP}, α_{TP}, α_{DP}, and γ,
Subcomplex III: β_{DP}, α_{DP}, α_E, and γ.

The packing structure of the $\alpha_3\beta_3\gamma$ complex can be assessed by calculating S for the three subcomplexes. Smaller $|S|$ implies that overall, the atoms in the subcomplex are more closely packed. S is calculated by the hybrid method of the angle-dependent integral equation (ADIE) theory [14–18] combined with a multipolar water model [15] and our morphometric approach (MA) [19–21]. The multipolar model is one of the most reliable molecular models for water [17, 18], and the hybrid method of the ADIE theory and the MA enables us to calculate the hydration entropy of a large polyatomic solute with sufficient accuracy and very high speed (see Chaps. 5 and 6 for more details). The three subcomplexes are named in terms of their positions. For example, when the γ subunit rotates by 120°, subcomplex III now comprises β_E, α_E, α_{TP}, and γ.

Each subcomplex from which the γ subunit is removed is referred to as "subcomplex I$-\gamma$", "subcomplex II$-\gamma$", or "subcomplex III$-\gamma$". Here, "$-$ (minus)" signifies the removal of the γ subunit. That is, the $\alpha_3\beta_3$ complex is decomposed into three subcomplexes as follows (see Fig. 3.10):

Subcomplex I$-\gamma$: β_E, α_E, and α_{TP},
Subcomplex II$-\gamma$: β_{TP}, α_{TP}, and α_{DP},
Subcomplex III$-\gamma$: β_{DP}, α_{DP}, and α_E.

The three subcomplexes are named in terms of their positions. For example, when the γ subunit rotates by 120°, subcomplexes III$-\gamma$ now comprises β_E, α_E, and α_{TP}. The decomposition described above is for analyzing the packing structure of the $\alpha_3\beta_3$ complex. We also calculate the packing efficiency of β_E, β_{TP}, or β_{DP}. The key quantity is S. Smaller $|S|$ implies that the atoms in the β subunit are more closely packed (i.e., the packing efficiency of the atoms in the β subunit is higher). Our primary concern is to verify that the order of packing efficiencies of subcomplexes I$-\gamma$, II$-\gamma$, and III$-\gamma$ is determined by that of β_E, β_{TP}, and β_{DP}.

We calculate the water-entropy gain upon contact of subunits X and Y, ΔS_{XY}, which is given by

$$\Delta S_{XY} = \text{"}S \text{ of subunit pair } X - Y\text{"} - (\text{"}S \text{ of subunit } X\text{"} + S \text{ of "subunit } Y\text{"}).$$

(3.1)

ΔS_{XY} is positive, and larger ΔS_{XY} implies that the atoms in the interface between subunits X and Y are more closely packed. The subunit pair $X-Y$ is taken from the complex, and subunits X and Y are obtained by simply separating the pair.

3.3.3 Results of Theoretical Analyses

Values of $\Delta S_{XY}/k_B$ (k_B is the Boltzmann constant) calculated for all the $\alpha-\beta$, $\alpha-\gamma$ and $\beta-\gamma$ pairs are collected in Table 3.1. In this table, $\Delta S_{XY}/k_B = 514.0$ for $X = \alpha_{DP}$ and $Y = \beta_{DP}$ implies that the water entropy increases by $514.0 k_B$ upon the contact of α_{DP} and β_{DP}. This increase arises primarily from the overlap of EVs generated by α_{DP} and β_{DP}. The contact produces the close packing of atoms in the $\alpha_{DP}-\beta_{DP}$ interface (i.e., the contact of atoms in α_{DP} and those in β_{DP} with the shape complementarity at the atomic level), leading to such a large water-entropy gain. The $\alpha_{DP}-\beta_{DP}$, $\alpha_E-\gamma$, and $\beta_{DP}-\gamma$ interfaces are the most closely packed and the $\alpha_{TP}-\beta_E$, $\alpha_{DP}-\gamma$, and $\beta_{TP}-\gamma$ interfaces are the most loosely packed among the $\alpha-\beta$, $\alpha-\gamma$ and $\beta-\gamma$ interfaces, respectively. The packing in the $\alpha_{TP}-\beta_{TP}$ and $\beta_E-\gamma$ interfaces is rather close, and

Table 3.1 Values of $\Delta S_{XY}/k_B$ (k_B is the Boltzmann constant) calculated for $\alpha-\beta$, $\alpha-\gamma$ and $\beta-\gamma$ pairs

Subunit Pair ($X-Y$)	$\Delta S_{XY}/k_B$
$\alpha_{DP}-\beta_{DP}$	514.0
$\alpha_{TP}-\beta_{TP}$	381.9
$\alpha_{DP}-\beta_{TP}$	291.9
$\alpha_E-\beta_{DP}$	283.8
$\alpha_E-\beta_E$	230.0
$\alpha_{TP}-\beta_E$	199.9
$\alpha_E-\gamma$	68.5
$\alpha_{TP}-\gamma$	17.5
$\alpha_{DP}-\gamma$	4.5
$\beta_{DP}-\gamma$	88.5
$\beta_E-\gamma$	65.4
$\beta_{TP}-\gamma$	37.8

the packing in the $\alpha_E-\beta_E$ and $\alpha_{TP}-\gamma$ interfaces is somewhat loose. These results are in good accord with the results from the molecular dynamics (MD) simulation with all-atom potentials by Ito and Ikeguchi [10, 13].

Values of S/k_B of subcomplexes I, II, and III are given in Table 3.2(a). The intra-subunit and inter-subunit contributions to S/k_B of each subcomplex are also given. The intra-subunit contribution is the sum of values of S/k_B for the subunits forming each subcomplex (e.g., β_E, α_E, α_{TP}, and the γ subunit forming subcomplex I). "S/k_B of each subcomplex" minus "intra-subunit contribution" is the inter-subunit contribution which represents the contribution from the interface packing between subunits in the subcomplex. A larger inter-subunit contribution implies higher packing efficiency in the interfaces. In Table 3.2(a), value for a subcomplex relative to that for subcomplex III is given in parentheses. It is observed that the values of S for the three subcomplexes follow the order, "$|S|$ of subcomplex III" $<<$ "$|S|$ of subcomplex II" $<$ "$|S|$ of subcomplex I". "$|S|$ of subcomplex III" including the $\alpha_{DP}-\beta_{DP}$, $\alpha_E-\gamma$, and $\beta_{DP}-\gamma$ interfaces is the smallest, indicating that the atoms in this complex is the most closely packed. Looking at the values in parentheses, we can conclude that the difference between subcomplexes in terms of $|S|$ comes primarily from that in terms of the inter-subunit contribution. Hereafter, we state that subcomplexes I, II, and III are loosely, moderately, and closely packed, respectively.

Table 3.2(b) gives values of S/k_B of subcomplexes I$-\gamma$, II$-\gamma$, and III$-\gamma$. Value for a subcomplex relative to that for subcomplex III is given in parentheses. We find that "$|S|$ of subcomplex III$-\gamma$" $<$ "$|S|$ of subcomplex II$-\gamma$" $<$ "$|S|$ of subcomplex I$-\gamma$". We can state that subcomplexes I$-\gamma$, II$-\gamma$, and III$-\gamma$ are loosely, moderately, and closely packed, respectively. The basic characteristics of the packing structure of the $\alpha_3\beta_3\gamma$ complex are determined by those of the $\alpha_3\beta_3$ complex. However, values of $|S|$ and the differences in terms of $|S|$ among subcomplexes I, II, and III are much larger than those among subcomplexes I$-\gamma$, II$-\gamma$, and III$-\gamma$. This result suggests that incorporation of the γ subunit in the $\alpha_3\beta_3$ complex enlarges the water-entropy

Table 3.2 Values of S/k_B (k_B is the Boltzmann constant) for subcomplexes I, II, and III (a) and for subcomplexes I$-\gamma$, II$-\gamma$, and III$-\gamma$ (b) defined in Fig. 3.10

(a)

Subcomplex	S/k_B	Intra-subunit contribution	Inter-subunit contribution
I	−61244.9 (−625.8)	−61826.8 (−234.6)	581.9 (−391.2)
II	−61027.9 (−408.8)	−61760.7 (−168.5)	732.8 (−240.3)
III	−60619.1 (0)	−61592.2 (0)	973.1 (0)

(b)

Subcomplex	S/k_B
I$-\gamma$	−50908.5 (−602.8)
II$-\gamma$	−50598.2 (−292.5)
III$-\gamma$	−50305.7 (0)

effect: In other words, the water entropy is quite sensitive to the orientation of the γ subunit in response to the packing structure of the $\alpha_3\beta_3$ complex.

An important result is that values of S/k_B of β_{DP}, β_{TP}, and β_E are -16390.0, -16407.4, and -16427.6, respectively. Value relative to that for β_{DP} is -17.4 for β_{TP} and -37.6 for β_E. Hence, "|S| of β_{DP}" < "|S| of β_{TP}" < "|S| of β_E". This order is reflected on the order, "|S| of subcomplex III−γ" < "|S| of subcomplex II−γ" < "|S| of subcomplex I−γ". The order, "|S| of β_{DP}" < "|S| of β_E" and "|S| of β_{TP}" < "|S| of β_E", coincides with the experimentally known information that β_{DP} and β_{TP} take closed structure but β_E takes open structure. "|S| of β_{DP}" < "|S| of β_{TP}", or equivalently, the feature that β_{DP} is more closely packed than β_{TP} (i.e., the structure of β_{DP} is more closed), can be unveiled only by our theoretical analyses.

3.3.4   Packing Structure Stabilized by Water-Entropy Effect

Figure 3.11 shows the packing structure stabilized in terms of the water entropy, which is revealed by our theoretical analyses. Subcomplexes I, II, and III are, respectively, loosely, moderately, and closely packed. The $\alpha_{DP}−\beta_{DP}$, $\alpha_E−\gamma$, and $\beta_{DP}−\gamma$ interfaces are the most closely packed among the α−β, α−γ and β−γ interfaces, respectively. The sign of high inequality "≪" appears only in the order, "|S| of subcomplex III" ≪ "|S| of subcomplex II". This is ascribed to the feature that the $\beta_{DP}−\gamma$ interface is more closely packed than the $\beta_{TP}−\gamma$ interface.

We define the orientation of the γ subunit as follows (see Fig. 3.12). Residues Arg8−Ile19 in the γ subunit come in contact with residues Asp386−Leu391 in

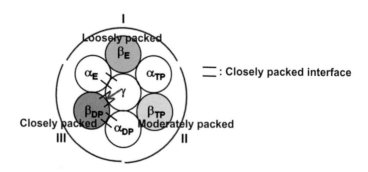

Fig. 3.11 Packing structure of $\alpha_3\beta_3\gamma$ complex in catalytic dwell state. The definitions of subcomplexes I, II, III, I−γ, II−γ, and III−γ are shown in Fig. 3.10. The orientation of the γ subunit is defined as indicated by the orange arrow. $S < 0$ is the hydration entropy. Smaller |S| implies a higher packing efficiency of the atoms in a protein or protein complex

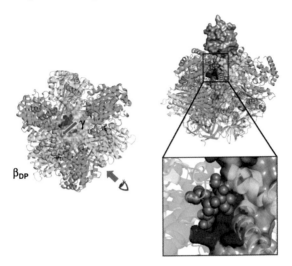

Fig. 3.12 Closely packed interface between β_{DP} and γ subunit and definition of orientation of γ subunit. Left: The orientation of the γ subunit is indicated by the orange arrow. Asp386-Leu391 in β_{DP} and Arg8-Ile19 in the γ subunit are represented by the orange and blue spheres, respectively. Right top: This is the picture viewed from the angle indicated by the gray arrow. Asp386-Leu391 in β_{DP} and Arg8-Ile19 in the γ subunit are represented by the orange and blue spheres, respectively. Right bottom: The portion surrounded by the red broken line is magnified. The convex surface of Asp386-Leu391 in β_{DP} and the concave surface of Arg8-Ile19 in the γ subunit are emphasized. See Fig. 3.3 also

β_{DP}. The former and the latter form portions with concave and convex surfaces, respectively. We note that the reduction of the EV upon the contact of concave and convex surfaces is considerably larger than that upon the usual contact of two convex surfaces as illustrated in Fig. 3.13. Moreover, close packing at the atomic level is attained between these two portions. As a consequence, the total EV is significantly reduced, leading to relatively large stabilization in terms of the water entropy. It was also found by Ito and Ikeguchi using the MD simulation with all-atom potentials [10, 13] that the contact between residues Arg8−Ile19 in the γ subunit and residues Asp386−Leu391 in β_{DP} is significantly stable. The orientation of the γ subunit achieving the $\beta_{DP}-\gamma$ interface possessing the aforementioned characteristics is emphasized by the vector indicated by the orange arrow in Fig. 3.12. In other words, the orientation is defined by this vector. We note, however, that the $\alpha_E-\gamma$ interface as well as the $\beta_{DP}-\gamma$ interface is closely packed and an important contributor to the stabilization of the particular orientation of the γ subunit.

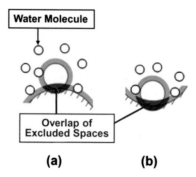

(a)   **(b)**

Fig. 3.13 a Contact of two convex surfaces. **b** Contact of convex and concave surfaces. The large spheres in (**a**) and (**b**) share the same diameter. The volume of the overlapping excluded space marked in black in (**b**) is larger than that in (**a**). When the convex surface of Asp386-Leu391 in β$_{DP}$ contacts the concave surface of Arg8-Ile19 in the γ subunit as illustrated in Fig. 3.12, a significantly large gain of water entropy is achieved

3.3.5   Relation Between Chemical Compound Bound and Packing Efficiency in a β Subunit

The order of packing efficiency for the β subunit depicted in Fig. 3.8 can now be made more specific. First, our theoretical analyses explained in Sects. 3.3.2 through 3.3.4 suggest that the packing efficiency of the β subunit follows the order, "β subunit with ATP• • •H$_2$O bound" > "β subunit with ATP bound" > "β subunit with Pi bound". Second, AMP-PNP is no more considered as a chemical compound bound to the β subunit. Third, we can construct a reasonable physical picture of the rotational mechanism described in Sect. 3.4 by further assuming that the packing efficiency follows the order, "β subunit with ATP" > "β subunit with ADP + Pi bound" and "β subunit with nothing bound" > "β subunit with Pi bound". As a consequence, we suggest that the packing efficiency of the β subunit follows the order depicted in Fig. 3.14.

3.4   Normal Rotation Under Solution Condition that ATP Hydrolysis Reaction Occurs: Rotation Mechanism

3.4.1   Basic Concept of Rotation Mechanism

The basic concept for constructing our physical picture of the rotational mechanism is summarized below.

(1) The packing efficiency of a β subunit is quite variable depending on the chemical compound bound to it (see Fig. 3.14).

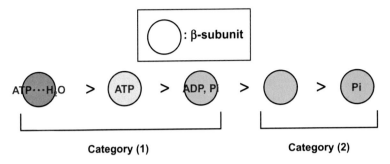

Category (1) **Category (2)**

Fig. 3.14 Order of packing efficiency for β subunit to which ATP•••H_2O (see the caption for Fig. 3.6), ATP, ADP + Pi, nothing, or Pi is bound. For example, the packing efficiency (PE) of a β subunit to which ATP•••H_2O is bound is higher than the PE to which ATP is bound and much higher than the PE to which Pi is bound. We also define ATP(ATP•••H_2O) (not shown here) as an intermediate between ATP and ATP•••H_2O. The PE of the β subunit with ATP(ATP•••H_2O) bound is lower than that with ATP•••H_2O bound but higher than that with ATP bound. Categories (1) and (2) are defined as follows: A β subunit to which ATP•••H_2O, ATP, or ADP + Pi is bound falls into category (1) whereas a β subunit to which nothing or Pi is bound falls into category (2)

(2) The packing efficiency of subcomplex I−γ, II−γ, or III−γ (i.e., the packing structure of the $\alpha_3\beta_3$ complex) is determined in accordance with that of the β subunit included in it. That is, as corroborated by our theoretical analyses, when the packing efficiency for a β subunit follows the order, "β subunit in subcomplex I−γ" < "β subunit in subcomplex II−γ" < "β subunit in subcomplex III−γ", the packing efficiency for a subcomplex follows the order, "subcomplex I−γ" < "subcomplex II−γ" < "subcomplex III−γ".

(3) The orientation of the γ subunit is optimized in response to the packing structure of the $\alpha_3\beta_3$ complex. The retainment of the highly stabilized β_{DP}−γ and α_E−γ contacts is especially important.

The packing structure of the $\alpha_3\beta_3\gamma$ complex possessing the specific nonuniformity shown in Fig. 3.11 is highly stable in terms of the water entropy. Here, we define categories (1) and (2) as indicated in Fig. 3.14. A β subunit to which ATP•••H_2O, ATP, or ADP + Pi is bound falls into category (1) whereas a β subunit to which nothing or Pi is bound falls into category (2). For the retainment of the high stability, at least it is required that two of the three β subunits be in category (1) and the other β subunit be in category (2). The $\alpha_3\beta_3\gamma$ complex is involved in the ATP hydrolysis cycle during which the chemical compound bound to each β subunit (i.e., ATP•••H_2O, ATP, ADP + Pi, nothing, or Pi) successively changes. Since the packing efficiency of a β subunit is strongly dependent on the chemical compound bound to it as depicted in Fig. 3.14, the packing efficiency of each β subunit successively changes during the ATP hydrolysis cycle, causing a successive change in the packing efficiency of each subcomplex (subcomplex I−γ, II−γ, or III−γ). The most important matter is to retain the specific nonuniformity of the packing structure mentioned above.

Subcomplex III$-\gamma$ possesses the highest packing efficiency. During the ATP hydrolysis cycle, the packing of subcomplex III$-\gamma$ becomes less efficient because the chemical compound bound to the β subunit in this subcomplex changes from ATP$\bullet \bullet$ H$_2$O to ADP + Pi due to the ATP hydrolysis. Subcomplex I$-\gamma$ is characterized by the lowest packing efficiency. During the ATP hydrolysis cycle, the packing of subcomplex I becomes more efficient because the chemical compound bound to the β subunit in this subcomplex changes from Pi to nothing due to the dissociation of Pi. The chemical compound bound to the β subunit in subcomplex II changes from ATP toward ATP$\bullet \bullet$ H$_2$O, leading to its higher packing efficiency. As explained in Sect. 3.4.2, in one ATP hydrolysis cycle, subcomplexes III$-\gamma$, II$-\gamma$, and I$-\gamma$ become loosely, closely, and moderately packed, respectively, resulting in a structural rotation of the $\alpha_3\beta_3$ complex by 120° in the counterclockwise direction. In response to the structural rotation of the $\alpha_3\beta_3$ complex, the γ subunit rotates by 120° in the same direction. Especially, it is crucial to recover the closely packed interfaces between the γ subunit and the β subunit to which ATP$\bullet \bullet \bullet$H$_2$O is bound, the β subunit named β_{DP}, and between the γ subunit and the α subunit adjacent to β_{DP} in the counterclockwise direction, the α subunit named α_{DP}.

3.4.2 Details of Rotation Mechanism

First, we summarize the experimentally available information on the rotational behavior [9, 22–25] (see Fig. 3.15).

(1) The ATP hydrolysis occurs in β_{DP} and Pi dissociates from β_E, with the result that the structure of β_{DP} becomes half-open [9]. This structural change of β_{DP} leads to a 40° rotation of the γ subunit [22–25]. The β subunit with this half-open structure is denoted by β_{DP}^{HO}. β_{TP} and β_E are renamed β'_{TP} and β'_E, respectively. ADP + Pi are bound to β_{DP}^{HO} and nothing is bound to β'_E. The chemical compound bound to β'_{TP} is ATP(ATP$\bullet \bullet \bullet$H$_2$O). Here, ATP(ATP$\bullet \bullet \bullet$H$_2$O) represents an intermediate between ATP and ATP$\bullet \bullet \bullet$H$_2$O.

(2) ADP dissociates from β_{DP}^{HO} and ATP binds to β'_E, inducing an 80° rotation of the γ subunit [22–25]. Changes of $\beta_{DP}^{HO}\rightarrow\beta_E$, $\beta'_{TP}\rightarrow\beta_{DP}$, and $\beta'_E\rightarrow\beta_{TP}$ occur. The $\alpha_3\beta_3\gamma$ complex now takes the structure that is the same as the structure before the 40° rotation of the γ subunit.

We then discuss the rotation mechanism on the basis of the information summarized above and the basic concept explained in Sect. 3.4.1 by starting from the catalytic dwell state shown as state (a) in Fig. 3.16. The ATP concentration is sufficiently high and the ADP and Pi concentrations are sufficiently low (see Sect. 2.1.1).

(1) The events which spontaneously occur in β_{DP}, β_E, β_{TP} are the ATP hydrolysis, dissociation of Pi, and change of ATP toward ATP$\bullet \bullet$H$_2$O (ATP$\rightarrow$ATP(ATP$\bullet \bullet$ $\bullet$H$_2$O)), respectively. Each event leads to a decrease in system free energy under

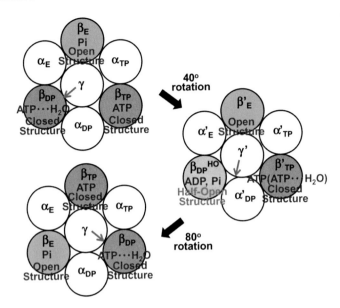

Fig. 3.15 Experimentally observed behavior of $\alpha_3\beta_3\gamma$ complex during one ATP hydrolysis cycle. The orientation of the γ subunit defined in Fig. 3.8 is employed. ATP(ATP• • •H_2O) represents an intermediate between ATP and ATP• • •H_2O (see the caption for Fig. 3.6)

the solution condition assumed. The packing efficiency of β_{DP} is lowered by the ATP hydrolysis, ATP• • •$H_2O \rightarrow$ ADP + Pi (see Fig. 3.14). That is, the structure of β_{DP} becomes half-open (i.e., looser), and the β subunit with this half-open structure is denoted by β_{DP}^{HO}. The packing efficiency of subcomplex III$-\gamma$, which is influenced dominantly by that of β_{DP}, is also lowered. Pi dissociates from β_E, and the packing of β_E and subcomplex I$-\gamma$ becomes closer (see Fig. 3.14). The resultant β subunit is renamed β'_E. Since ATP in β_{TP} changes toward ATP• • •H_2O, the packing efficiency of β_{TP} becomes higher (i.e., the packing of β_{TP} and subcomplex II$-\gamma$ becomes closer) (see Fig. 3.14). The resultant β subunit is renamed β'_{TP}.

(2) The $\alpha_3\beta_3\gamma$ complex is now in state (b) shown in Fig. 3.16. In state (b), the packing of β_{DP}^{HO} with ADP + Pi bound is looser than that of β'_{TP} with ATP(ATP• • •H_2O) bound (see Fig. 3.14). Here, the packing efficiency of the β subunit with ATP(ATP• • •H_2O) bound is lower than that with ATP• • •H_2O bound but higher than that with ATP bound. ADP dissociates from β_{DP}^{HO}, giving rise to lower packing efficiency of this β subunit (see Fig. 3.14). The packing of β_{DP}^{HO} and subcomplex III$-\gamma$ becomes looser. ATP binds to β'_E, making the packing efficiency of this β subunit higher (see Fig. 3.14). The packing of β'_E and subcomplex I$-\gamma$ becomes closer. The change which can be written as "ATP(ATP• • •$H_2O) \rightarrow$ (ATP• • •H_2O)" (further change of ATP toward ATP• • •H_2O) occurs in β'_{TP} with the result of higher packing efficiency of this β subunit (see Fig. 3.14). The packing of β'_{TP} and subcomplex II$-\gamma$ becomes closer.

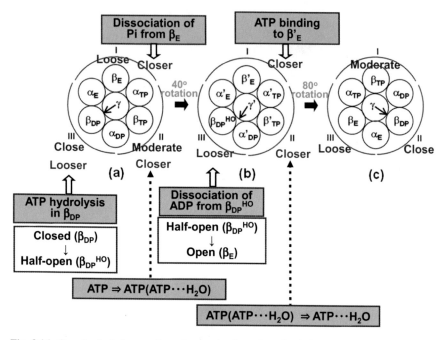

Fig. 3.16 Our physical picture of rotation mechanism of γ subunit in $\alpha_3\beta_3\gamma$ complex during one ATP hydrolysis cycle in scenario (A). "ATP• • •H₂O" represents ATP just before the hydrolysis reaction (the activated complex). ATP(ATP• • •H₂O) represents an intermediate between ATP and ATP• • •H₂O. The packing of β_E (and subcomplex I−γ) is loose in state (a) but it becomes closer in state change (a)→(b). The packing of β'_E (and subcomplex I−γ) becomes further closer in (b)→(c). The packing of β_{TP} (and subcomplex I−γ) is now moderate in (c). The packing of β_{TP} (and subcomplex II−γ) is moderate in (a) but it becomes closer in (a)→(b). The packing of β'_{TP} (and subcomplex II−γ) becomes further closer in (b)→(c). The packing of β_{DP} (and subcomplex III−γ) is now close in (c). The packing of β_{DP} (and subcomplex III−γ) is close in (a) but it becomes looser in (a)→(b). The packing of β_{DP}^{HO} (and subcomplex III−γ) becomes further looser in (b)→(c). The packing of β_E (and subcomplex III−γ) is now loose in (c)

(3) Thus, changes of $\beta_{DP}\rightarrow\beta_{DP}^{HO}\rightarrow\beta_E$, $\beta_E\rightarrow\beta'_E\rightarrow\beta_{TP}$, and $\beta_{TP}\rightarrow\beta'_{TP}\rightarrow\beta_{DP}$ occur in subcomplexes I−γ, II−γ, and III−γ, respectively. The γ subunit rotates by 120° in the counterclockwise direction in response to the change in the packing structure of the $\alpha_3\beta_3$ complex, primarily to recover the closely packed interfaces with β_{DP} and α_E. The $\alpha_3\beta_3\gamma$ complex is now in state (c) shown in Fig. 3.16. Figure 3.17 depicts how the chemical compound bound to the β subunit in each of subcomplexes I−γ, II−γ, and III−γ changes during one ATP hydrolysis cycle. The information on the resulting change in packing efficiency of the β subunit is also provided in Fig. 3.17. Though the overall structure before and after the 120° rotation are the same, one ATP molecule is hydrolyzed in aqueous solution: The system free energy becomes lower by the free-energy change upon the APT hydrolysis reaction, $\Delta G \sim -20k_BT$ ($T = 298$ K) (see Sect. 2.1.1).

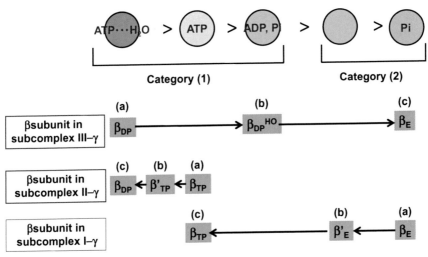

Fig. 3.17 Changes in chemical compounds bound to three β subunits in subcomplexes I−γ, II−γ, and III−γ, respectively, during one ATP hydrolysis cycle in scenario (A). In state (a), for example, ATP• • •H_2O (see the caption for Fig. 3.6), ATP, and Pi are bound to β_{DP}, β_{TP}, and β_E, respectively. ATP(ATP• • •H_2O), an intermediate between ATP and ATP• • •H_2O, is bound to β'_{TP} in state (b). ATP(ATP• • •H_2O) represents an intermediate between ATP and ATP• • •H_2O. The packing efficiency (PE) of β'_{TP} in state (b) is lower than the PE of β_{DP} in state (a) or (c) but higher than the PE of β_{TP} in state (a) or (c). In states (a), (b), and (c), two of the three β subunits are in category (2) and the other β subunit is in category (1) (see Fig. 3.14), thus retaining the high stability of the $\alpha_3\beta_3\gamma$ complex

In state (a) shown in Fig. 3.16, the ATP dissociation from β_{TP}, for example, is not likely to occur for the following two reasons: First, under the solution condition assumed, the ATP dissociation causes an increase in system free energy; second, in the resultant state, two of the three β subunits are in category (2) and the other β subunit is in category (1), vitiating the high structural stability of the $\alpha_3\beta_3\gamma$ complex.

The nonuniform packing structure of state (a), which is most favored by the water-entropy effect, is thus retained during one ATP hydrolysis cycle for preventing a water-entropy loss. Currently, it is difficult to explain why the γ subunit rotates by 40° in state change (a)→(b) and by 80° in state change (a)→(b). If the crystal structure of state (b) was experimentally available, the explanation would be made possible by performing additional theoretical analyses.

3.4.3 Crucial Importance of Water-Entropy Effect in Unidirectional Rotation

A protein complex is driven to take a structure which mitigates the water crowding and increases the translational, configurational entropy of water as much as possible: This is the water-entropy effect (see Sect. 2.8). The β subunits in F_1-ATPase are coupled with the ATP hydrolysis reaction, an irreversible process, through their catalytic actions. They are involved in the ATP hydrolysis cycle. Due to the water-entropy effect, the packing efficiencies of the three β subunits and those of subcomplexes I−γ, II−γ, and III−γ become significantly different. The orientation of the γ subunit is determined in response to the nonuniform packing structure of the $\alpha_3\beta_3$ complex through the water-entropy effect. The retainment of the closely packed β_{DP}−γ and α_E−γ interfaces is particularly important. The packing efficiency of a β subunit is intimately related to the chemical compound bound to it. As the ATP hydrolysis reaction proceeds, the chemical compounds bound to the three β subunits successively change, and the packing efficiencies of the three β subunits and those of subcomplexes I−γ, II−γ, and III−γ also successively change. This structural change of the $\alpha_3\beta_3$ complex induces the orientational change of the γ subunit.

The hydration entropy of each β subunit to which one of ATP, ADP + Pi, Pi, and nothing is bound, that of the $\alpha_3\beta_3$ complex, and the large dependence of hydration entropy of the $\alpha_3\beta_3\gamma$ complex on the orientation of the γ subunit are important thermodynamic quantities for elucidating the unidirectional rotation. The force for rotating the γ subunit is generated by not ATP but water. If the rotation was driven by the free energy by ATP hydrolysis reaction as claimed by the prevailing view, the γ subunit would rotate by 120° during the hydrolysis in β_{DP}.

After the 120° rotation of the γ subunit, the absolute value of hydration entropy |S| of a subcomplex (the magnitude of water-entropy loss caused by the presence a subcomplex) increases by ~$626k_B$ for subcomplex III, decreases by ~$−409k_B$ for subcomplex II, and decreases by ~$−217k_B$ for subcomplex I (see Table 3.2(a)). Therefore, the hydration entropy of the $\alpha_3\beta_3\gamma$ complex is kept almost unchanged during one ATP hydrolysis cycle, i.e., a single rotation (the 120° rotation). During the rotation, a very large increase in |S| for one of the three subcomplexes is cancelled out by large decreases in |S| for the other two subcomplexes, causing no free-energy barrier for the rotation of the γ subunit.

It was shown with surprise by an experimental study [1] that the three β subunits in the $\alpha_3\beta_3$ complex (i.e., without the γ subunit) exhibit cyclic structural changes in the same rotary direction as in the $\alpha_3\beta_3\gamma$ complex (see Fig. 3.5). Only one of the three β subunits takes open structure, and when an open-to-closed transition occurs for this β subunit, the opposite closed-to-open transition simultaneously occurs for its counterclockwise neighboring β subunit. This experimental result is in accord with our physical picture discussed in Sect. 3.4.2.

Geometrically, the axle of the γ subunit penetrates the central cavity with cylindrical shape of the $\alpha_3\beta_3$ complex (see Fig. 3.12). In an interesting experimental work [26], the γ subunit was truncated so that the remaining head of the γ subunit outside the cavity could sit on the concave entrance of the $\alpha_3\beta_3$ complex. Strikingly, the truncated γ subunit rotated in the normal direction, though the average rotary speed was lower and moments of irregular motion were exhibited. This experimental result is suggestive of the following: Wide, closely packed α−γ and β−γ interfaces are not necessarily required for entropically correlating the γ subunit with the packing structure of the $\alpha_3\beta_3$ complex; even narrow, closely packed interfaces enable the γ subunit to rotate in the normal direction in response to the packing structure of the $\alpha_3\beta_3$ complex; however, such narrow interfaces weakens the entropic force by water, causing the slower and more irregular motion of the γ subunit (see Sect. 3.8.4 also).

3.4.4 Change in System Free Energy During a Single Rotation

After the 120° rotation of the γ subunit, the $\alpha_3\beta_3\gamma$ complex returns to the same state, but one ATP molecule is decomposed into ADP and Pi by the hydrolysis reaction (ATP + $H_2O \rightarrow$ ADP + Pi). As depicted in Fig. 3.18, the dissociation of Pi and the ATP hydrolysis occur in state change (a)→(b), and the ATP binding and the dissociation of ADP occur in state change (b)→(c). The structure of the $\alpha_3\beta_3\gamma$ complex is reorganized in each of state changes (a)→(b) and (b)→(c) to retain the

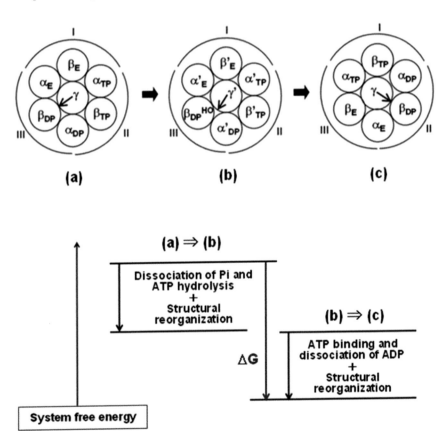

Fig. 3.18 Decrease of system free energy during a single rotation of γ subunit (i.e., one ATP hydrolysis cycle)

water entropy already maximized as discussed in Sect. 3.4.2. The system free energy decreases by ΔG_{ab} and ΔG_{bc} in state changes (a)→(b) and (b)→(c), respectively, and $\Delta G_{ab} + \Delta G_{bc} = \Delta G \sim -20k_B T$ ($T = 298$ K).

3.4.5 Effect of Electrostatic Attractive Interaction Between γ and β Subunits

For a protein or protein complex, a decrease in intramolecular energy by van der Waals and electrostatic attractive interactions is unavoidably accompanied by a loss of protein-water van der Waals and electrostatic attractive interactions. The loss causes an energy increase referred to as the energetic dehydration penalty explained in Sect. 2.6. The decrease in the intramolecular energy, which is almost cancelled out by the energetic dehydration penalty, cannot be a driving force of the rotation of the γ subunit.

It was suggested that the rotation was induced by the electrostatic, attractive interaction between positively charged residues (Arg and Lys) on the protruding portion of the γ subunit and negatively charged residues (Asp (D) and Glu (E)) in the so-called DELSEED motif of the β subunit (corresponding to Asp394−Asp400 in the $\alpha_3\beta_3\gamma$ complex being considered) [27]. However, it was shown in later experimental works [28, 29] that the rotation is not influenced by mutating each residue and all five acidic residues in the DELSEED motif to Ala (the kinetic parameters are comparable to those of the wild type), demonstrating that the electrostatic, attractive interaction mentioned above plays no roles for the rotation.

3.5 Theoretical Analyses Based on Experimental Observations for Yeast F_1-ATPase

Unfortunately, no crystal structures have been reported for state (b) shown in Fig. 3.16. Instead, crystal structures are available for the catalytic dwell state of yeast F_1-ATPase and the state after $16°$ rotation of γ subunit [30]. In the catalytic dwell structure, AMP-PNP is bound to β_{DP} and β_{TP} and Pi is bound to β_E as illustrated in Fig. 3.19. The ATP hydrolysis does not occur in β_{DP} during the $16°$ rotation, and the rotation is triggered by the dissociation of Pi from β_E.

We analyze the packing structures of the two crystal structures mentioned above (see Fig. 3.19) and the packing efficiencies of the α−β, α−γ and β−γ interfaces. AMP-PNP is replaced by ATP. Refer to our earlier publication for more details [31]. The purpose of these analyses is to investigate the effect of Pi bound to β_E on the packing structure of the $\alpha_3\beta_3\gamma$ complex. It is interesting to know whether the subtle difference between the left and right states shown in Fig. 3.19 in the packing structure can be reproduced.

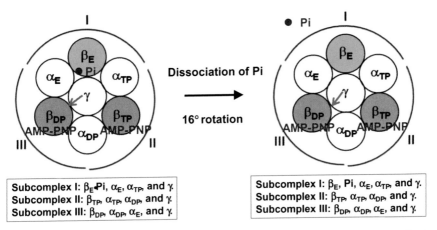

Fig. 3.19 States of yeast F$_1$-ATPase before and after 16° rotation of γ subunit. The 16° rotation is triggered by the dissociation of Pi from β$_E$

As illustrated in Fig. 3.19, not only subcomplexes II and III but also subcomplex I before the 16° rotation and that after the rotation are defined as

Subcomplex I before rotation: β$_E$·Pi, α$_E$, α$_{TP}$, and γ,
Subcomplex I after rotation: β$_E$, Pi, α$_E$, α$_{TP}$, and γ.

Here, β$_E$·Pi represents β$_E$ to which Pi is bound. ATP-Mg^{2+} is bound to β$_{TP}$ and β$_{DP}$ in the analyses. The calculation of the hydration entropy is performed by a hybrid of the ADIE theory [14–18] combined with a multipolar water model [15] and the MA [19–21].

We find that upon the rotation, the absolute value of hydration entropy |S| of a subcomplex (the magnitude of water-entropy loss caused by the insertion of a subcomplex) increases by ~30k_B for subcomplex III, decreases by ~−21k_B for subcomplex II, and decreases by ~−47k_B for subcomplex I. These values are not very accurate in a quantitative sense for the following reason: The two crystal structures before and after 16° rotation are not significantly different and we take differences between two large quantities in calculating it, giving rise to inevitable cancellation of significant digits. It is definite, however, that |S| increases for subcomplex III and |S| decreases for subcomplexes I and II though ATP• • •H$_2$O→ADP + Pi and ATP→ATP(ATP• • •H$_2$O) do not occur in β$_{DP}$ of subcomplex III and β$_{TP}$ of subcomplex II, respectively, unlike in Sects. 3.4.2 and 3.4.3. Importantly, an increase in |S| for one of the three subcomplexes is always compensated with decreases in |S| for the other two subcomplexes. This argument supports our picture of the rotation mechanism explained in Sects. 3.4.2 and 3.4.3. The increase in water entropy upon the rotation is ~38k_B (38k_B = −30 k_B + 21k_B + 47k_B). Thus, the increase in water entropy (~−38$k_B T$ when converted to the free-energy decrease) is comparable with (i.e., roughly as small as) the decrease in system free energy during one ATP hydrolysis cycle ~ −20k_BT ($T = 298$ K), which is physically reasonable. We also find that the interfaces of α$_{DP}$−β$_{DP}$ and α$_E$−β$_E$ become less efficiently packed (i.e., more open) after the rotation, which is in agreement with the experimental observations [30].

3.6 Inverse Rotation Under Solution Condition that ATP Synthesis Reaction Occurs

3.6.1 State of α$_3$β$_3$γ Complex Stabilized

We now discuss scenario (B) defined in Sect. 3.3. The ATP concentration is sufficiently low and the ADP and Pi concentrations are sufficiently high (see Sect. 2.1.1). The state stabilized is shown in Fig. 3.20: It is visually the same as the state shown in Fig. 3.9. However, ATP• • •H$_2$O in scenario (B) denotes ATP just after the ATP synthesis reaction. The states in Figs. 3.9 and 3.20 share the same packing structure shown in Fig. 3.11. The events occurring during one ATP synthesis cycle are the binding of ADP and Pi, ATP synthesis, and dissociation of ATP. Each event, which leads to a decrease in system free energy under the solution condition assumed, spontaneously occurs.

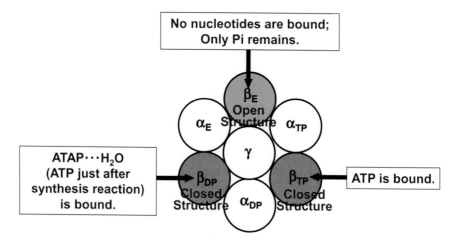

Fig. 3.20 Initial state of α$_3$β$_3$γ complex for one ATP synthesis cycle in scenario (B)

3.6.2 Details of Rotation Mechanism

We infer that the rotation mechanism is the following (see Fig. 3.21).

(1) ADP binds to β_E, ATP dissociates from β_{TP}, and "ATP• • •H_2O→ATP(ATP• • •H_2O)" (change of ATP• • •H_2O toward ATP) occurs in β_{DP} in state change (a)→(b). These three events lead to enhanced, lowered, and lowered packing efficiencies of β_E, β_{TP}, and β_{DP}, respectively (see Fig. 3.14), and the packings of subcomplexes I−γ, II−γ, and III−γ become closer, looser, and looser, respectively.

(2) In state change (b)→(c), ATP synthesis (ADP + Pi→ATP• • •H_2O) occurs in β'_E, Pi binds to β'_{TP}, and "ATP(ATP• • •H_2O)→ATP" (further change of ATP• • •H_2O toward ATP) occurs in β'_{DP}. These three events lead to enhanced, lowered, and lowered packing efficiencies of β'_E, β'_{TP}, and β'_{DP}, respectively (see Fig. 3.14), and the packings of subcomplexes I−γ, II−γ, and III−γ become closer, looser, and looser, respectively.

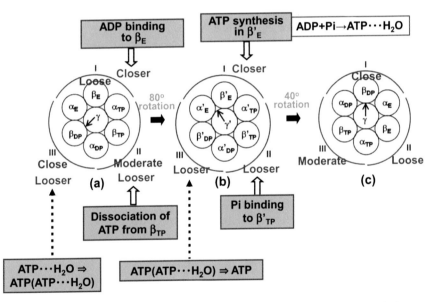

Fig. 3.21 Our physical picture of rotation mechanism of γ subunit in $\alpha_3\beta_3\gamma$ complex during one ATP synthesis cycle in scenario (B. "ATP• • •H_2O" represents ATP just after the synthesis reaction. ATP(ATP• • •H_2O) represents an intermediate between ATP and ATP• • •H_2O. The packing of β_E (and subcomplex I−γ) is loose in state (a) but it becomes closer in stage change (a)→(b). The packing of β'_E (and subcomplex I−γ) becomes further closer in (b)→(c). The packing of β_{TP} (and subcomplex II−γ) is moderate in (a) but it becomes looser in (a)→(b). The packing of β'_{TP} (and subcomplex II−γ) becomes further looser in (b)→(c). The packing of β_E (and subcomplex II−γ) is now loose in (c). The packing of β_{DP} (and subcomplex III−γ) is close in (a) but it becomes looser in (a)→(b). The packing of β'_{DP} (and subcomplex III−γ) becomes further looser in (b)→(c). The packing of β_{TP} (and subcomplex III−γ) is now moderate in (c)

(3) Thus, changes of $\beta_{DP} \rightarrow \beta'_{DP} \rightarrow \beta_{TP}$, $\beta_E \rightarrow \beta'_E \rightarrow \beta_{DP}$, and $\beta_{TP} \rightarrow \beta'_{TP} \rightarrow \beta_E$ occur in subcomplexes I−γ, II−γ, and III−γ, respectively. The γ subunit rotates by 120° in the clockwise direction in response to the change in the packing structure of the $\alpha_3\beta_3$ complex, primarily to recover the closely packed interfaces with β_{DP} and α_E. Though the overall structures before and after the 120° rotation are the same, one ATP molecule is synthesized in aqueous solution: The system free energy becomes lower by the free-energy change upon the APT synthesis reaction in aqueous solution under the physiological condition, $-\Delta G \sim -20k_B T$ ($T = 298$ K). We note that in the solution condition that the ATP synthesis reaction occurs, ΔG defined by Eq. (2.1) is positive (see Sect. 2.1.1).

3.7 Normal and Inverse Rotations with the Same Frequency (Rotations in Random Directions) Under Solution Condition that ATP Hydrolysis and Synthesis Reactions Are Equilibrated

In scenario (C) defined in Sect. 3.3, the ATP hydrolysis and synthesis reactions are in equilibrium. Even in this chemical equilibrium state, the reactions are not stopped: The hydrolysis and synthesis reactions occur with the same frequency (see Sect. 2.1.1). Hence, the ATP hydrolysis accompanied by the normal rotation expounded in Fig. 3.16 and the ATP synthesis leading to the inverse rotation expounded in Fig. 3.21 take place with the same frequency. States (a) in these two figures are the same, and the chemical compounds bound to β_{DP}, β_{TP}, and β_E are, respectively, ATP• • •H_2O, ATP, and Pi. In essence, the rotation does not occur.

Even under the solution condition that the ATP hydrolysis reaction occurs, the frequency of the ATP synthesis reaction does not vanish (see Sect. 2.1.1): It is just that the frequency of the ATP hydrolysis reaction is much higher than that of the ATP synthesis reaction. Hence, in scenario (A) the ATP synthesis reaction rarely occurs with the inverse rotation, which was actually observed in experiments [22].

3.8 Inverse Rotation Compelled by External Torque Imposed on Central Shaft and Occurrence of ATP Synthesis Under Solution Condition that ATP Hydrolysis Reaction Should Occur

3.8.1 What Will Happen When Inverse Rotation is Forcibly Executed?

We then discuss scenario (D) defined in Sect. 3.3. It was experimentally revealed by Muneyuki and coworkers [32–35] using F_1-ATPase from thermophilic *Bacillus* strain

PS3 that even under the solution condition that the ATP hydrolysis reaction should occur, the application of sufficiently strong external torque to the γ subunit using an advanced technique induces the inverse rotation accompanied by the occurrence of ATP synthesis reaction. This intriguing behavior can be explicated by the competition of the entropic force by water driving the normal rotation and the external force driving the inverse rotation.

As discussed in Sects. 3.3.2–3.3.4, for the packing structure of the $\alpha_3\beta_3$ complex given, the water entropy is strongly dependent on the orientation of the γ subunit. Without the application of the external torque, the γ subunit changes its orientation in accordance with the change in packing structure of the $\alpha_3\beta_3$ complex, which is induced by the ATP hydrolysis cycle. In the case where the inverse rotation is forcibly executed, on the other hand, the $\alpha_3\beta_3$ complex changes its packing structure in accordance with the orientational change of the γ-subunit. Importantly, the packing-structure change of the $\alpha_3\beta_3$ complex is accomplished by changing the chemical compounds bound to the three β subunits.

When the inverse rotation of the γ subunit is forcibly executed, state change (a)→(b) (states (a) and (b) are the same) illustrated in Fig. 3.22 occurs for the following reason. The hydration entropies of the $\alpha_3\beta_3\gamma$ complex in states (a) and (b) are the same, and state change (a)→(b) does not give rise to a water-entropy loss (factor 1). It should be emphasized that the solution is under the condition that the ATP hydrolysis reaction should occur. Hence, during state change (a)→(b), an increase in system free energy is caused by the ADP and Pi binding to, ATP synthesis in, and ATP dissociation from the $\alpha_3\beta_3\gamma$ complex (factor 2). In state change (a)→(c) or (a)→(d) illustrated in Fig. 3.23, on the other hand, an acceptably large loss of water entropy is unavoidable (factor 1). A decrease in system free energy is caused by the ATP binding to, ATP hydrolysis in, and ADP and Pi dissociation from the $\alpha_3\beta_3\gamma$ complex during state change (a)→(c), and no change in system free energy

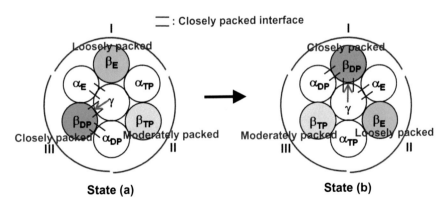

Fig. 3.22 Inverse rotation of γ subunit forcibly executed: state change (a)→(b) during which the ADP and Pi binding to, ATP synthesis in, and ATP dissociation from the $\alpha_3\beta_3\gamma$ complex take place (the ATP synthesis occurs). The water entropy remains constant (i.e., the water entropy already maximized is retained)

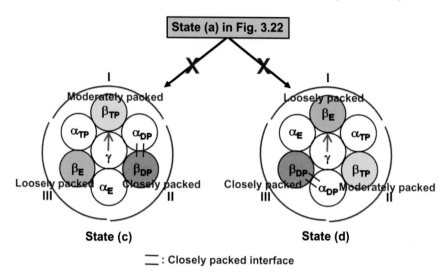

Fig. 3.23 Inverse rotation of γ subunit forcibly executed: state changes (a)→(c) and (a)→(d) (see Fig. 3.22 for state (a)) where the ATP hydrolysis and neither the hydrolysis nor synthesis occurs, respectively. During state change (a)→(c), the ATP binding to, ATP hydrolysis in, and ADP and Pi dissociation from the α$_3$β$_3$γ complex take place. An unacceptably large loss of water entropy is caused in state change (a)→(c) or (a)→(d)

is caused during state change (a)→(d) (factor 2). Factor 1 is much more important than factor 2 in the minimization of system free energy, and state change (a)→(b) (see Fig. 3.22) spontaneously occurs.

In summary, for retaining the water entropy already maximized, the chemical compounds bound to the three β subunits and the packing structure of the α$_3$β$_3$ complex are optimized in response to the orientation of the γ subunit. The chemical compounds bound to the three β subunits are updated during state change (a)→(b) as depicted in Fig. 3.24. Upon state change (a)→(b), the system free energy increases by $20k_BT$ ($T = 298$ K) in aqueous solution under the physiological condition (see Sect. 2.1.1). This is not contradictory to thermodynamics, because a large energy is given to the system through the external torque (See case III in Sect. 3.8.2 for more details).

3.8.2 Three Cases Where Normal Rotation Persists, Inverse Rotation Occurs, and Essentially no Rotations Occur When External Torque is Applied

We consider the case where the external torque is applied to the γ subunit under the solution condition that the ATP hydrolysis reaction should occur. The strength of external torque multiplied by $2\pi/3$ (120°) is the work performed on the γ subunit per

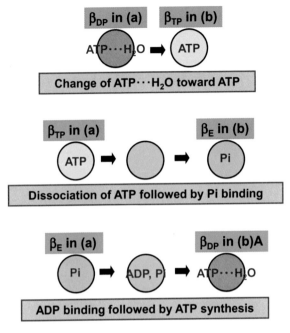

Fig. 3.24 Updates of chemical compounds bound to three β subunits during inverse rotation compelled by external torque imposed on γ subunit (i.e., state change (a)→(b) illustrated in Fig. 3.22). "ATP• • •H$_2$O" represents ATP just after the synthesis reaction

ATP cycle [34], W_{ext}. Here, the ATP hydrolysis, ATP synthesis, or neither hydrolysis nor synthesis occurs "per ATP cycle". When a portion of W_{ext}, Q, is lost as the heat, the work received by the system, W, is given by

$$W = W_{ext} - Q. \tag{3.2}$$

$W > 0$ is an energy given to the system per ATP cycle. In scenario (D), the change in system free energy per ATP cycle is $\Delta G + W$ where ΔG is given by Eq. (2.1). $\Delta G = -20k_B T$ ($T = 298$ K) in aqueous solution under the physiological condition.

The following three cases can be considered (see Sect. 2.1.1 also).

Case I. $\Delta G + W < 0$. The external force is weaker than the entropic force by water. The γ subunit rotates in the normal direction. More strictly, the frequency of the normal rotation is higher than that of the inverse rotation, and the normal rotation dominates when $\Delta G + W$ is not close to zero. The overall reaction occurring is the ATP hydrolysis, ATP + H$_2$O→ADP + Pi. On an average, the system free energy decreases by $\Delta G + W$ per ATP cycle.

Case II. $\Delta G + W = 0$. The external force is as strong as the entropic force by water. The rotation of the γ subunit occurs in random directions: The γ subunit rotates in the normal and inverse directions with the same frequency, and the rotation essentially vanishes. The ATP hydrolysis and synthesis reactions occur with the same frequency. On an average, the net change in system free energy per ATP cycle is zero.

Case III. $\Delta G + W > 0$. The external force is stronger than the entropic force by water. The γ subunit rotates in the inverse direction. More strictly, the frequency of the inverse rotation is higher than that of the normal rotation, and the inverse rotation dominates when $\Delta G + W$ is not close to zero. The overall reaction occurring is the ATP synthesis, ADP + Pi→ATP + H_2O. On an average, the system free energy increases by $\Delta G + W$ per ATP cycle. This is not thermodynamically inconsistent, because a large energy of W is given to the system.

3.8.3 Comparison with Experimental Results Observed When External Torque is Applied

It was experimentally observed in case I that as the external torque becomes stronger, the apparent rotation rate decreases [34]. This result suggests that the normal rotation occurs almost exclusively for sufficiently weak torque but the frequencies of no rotation and the inverse rotation increase as the torque becomes stronger. Let us assume that $\Delta G + W = -16k_BT$ ($T = 298$ K). Suppose that the aqueous solution is under the physiological condition and $\Delta G = -20k_BT$ ($T = 298$ K). Since states (a) and (b) shown in Fig. 3.22 share the same value of the hydration entropy, the free-energy balance can be considered as follows. The energy given to the system per ATP cycle is $W = 4k_BT$. Here, we introduce two possible occasions. In the first occasion, the normal and inverse rotations occur $9x/10$ times and $x/10$ times, respectively (x is a large, positive value). The normal and inverse rotations are accompanied by the ATP hydrolysis and synthesis, respectively. The change in system free energy upon the hydrolysis and synthesis are $-20k_BT$ and $20k_BT$, respectively. The free-energy balance is expressed as $\{(9x/10)(-20k_BT) + (x/10)(20k_BT)\}/x = -16k_BT$, which coincides with $\Delta G + W = -16k_BT$. That is, the system free energy decreases by $-16k_BT$ per ATP cycle on an average. In the second occasion, the normal rotation occurs $4x/5$ times and the rotation is stopped $x/5$ times. When the rotation is stopped, neither the ATP hydrolysis nor synthesis occurs with the system free energy being unchanged. The free-energy balance is expressed as $\{(4x/5)(-20k_BT) + (x/5)(0k_BT)\}/x = -16k_BT$. In the real system, all the normal, inverse, and stopped rotations should occur. When the torque is very weak, the normal rotation occurs almost exclusively. However, with an increase in torque strength, the proportion of the stopped rotation becomes higher. With a further increase in torque strength toward $\Delta G + W = 0$, the proportion of the inverse rotation also becomes higher.

In case II where the rotation essentially vanishes, the normal and inverse rotations occur with the same frequency, and there should be significantly many stopped rotations as well. In fact, it was experimentally observed that the frequency of the normal and inverse rotations is quite low in the case of $\Delta G + W = 0$ [34]. When the ATP, ADP, and Pi concentrations in the aqueous solution are changed (i.e., they are not under the physiological condition) but the solution is still under the condition that the ATP hydrolysis reaction should occur, $\Delta G \neq -20 k_B T$ ($T = 298$ K) but $\Delta G < 0$ is still true. Theoretically, $W = -\Delta G$ holds irrespective of the ATP, ADP, and Pi concentrations. In fact, Muneyuki and coworkers found that $W = -\Delta G$ holds at all the ATP, ADP, and Pi concentrations tested [34].

The experimental observations in case III indicated that as the external torque becomes stronger, the apparent rotation rate increases [34]. The free-energy balance can be discussed in a manner which is similar to that in case I. All the normal, inverse, and stopped rotations should occur. With an increase in torque strength, the proportion of the normal rotation becomes lower. With a further increase in torque strength, the proportion of the stopped rotation also becomes lower. Once the torque becomes sufficiently strong, the inverse rotation occurs almost exclusively.

3.8.4 Substantially Different Behavior Observed for a Mutant of F_1-ATPase

Muneyuki and coworkers [35] carried out interesting experiments using a mutant of F_1-ATPase. In this mutant, a mutation of E190D is made for every β subunit. The residue E190 is located near the γ-phosphate of ATP bound to the catalytic site and significantly affects the ATPase activity [36]. The cleavage of the $\beta-\gamma$-phosphate bond of ATP is markedly delayed by the mutation, resulting in the long catalytic dwell [24]. In what follows, we summarize the experimental results obtained for the mutant and comment on them.

(i)   The mutant cannot form a stabilized structure without the γ subunit [35]. This result suggests that it is not achievable to closely pack the $\alpha-\beta$ interfaces of the $\alpha_3\beta_3$ complex and the packing efficiencies of the three β subunits are not well reflected on those of subcomplexes I$-\gamma$, II$-\gamma$, and III$-\gamma$. In comparison with the wild type, the structural stability of the $\alpha_3\beta_3$ complex is less influenced by the chemical compound bound to each β subunit.

(ii) With the γ subunit, the mutant forms a stable complex [35]. The $\alpha_3\beta_3\gamma$ complex of the mutant is capable of hydrolyzing ATP into ADP and Pi with the rotation of the γ subunit in the normal direction under the solution condition that the ATP hydrolysis reaction occurs. However, the thermal denaturation temperature of the mutant is significantly lower than that of the wild type [35]. This result indicates that the $\alpha_3\beta_3\gamma$ complex of the mutant is less stable than that of the wild type. In the mutant, the orientation of the γ subunit is less correlated with the packing structure of the $\alpha_3\beta_3$ complex.

(iii) By the application of external torque to the γ subunit, the rotation essentially vanishes with an external torque which is considerably weaker than for the wild type [35]. Let W_{Wild} and W_{Mutant} be the values of W with which the rotation essentially vanishes for the wild type and the mutant, respectively. From the experimental result mentioned above, $W_{\text{Mutant}} < W_{\text{Wild}}$. Since the values of ΔG

for the mutant and the wild type are the same, $W_{Wild} = -\Delta G$ ($\Delta G + W_{Wild} = 0$) means $\Delta G + W_{Mutant} < 0$. Hence, even when the rotation essentially vanishes for the mutant, the overall reaction occurring is still the ATP hydrolysis. When the external torque is further increased but $\Delta G + W_{Mutant} < 0$ still holds (i.e., the torque is not very strong), the γ subunit rotates in the inverse direction but the overall reaction occurring is still the ATP hydrolysis.

(iv)   When the external torque is applied, the rotation becomes quite irregular and the rotation rate frequently exhibits an abrupt increase or decrease [35].

It is not straightforward to interpret the aforementioned results because no information on the mutant structure is experimentally available. However, we can give a speculative discussion in what follows.

The wild type fortuitously meets all the following requirements for the functional expression: (1) The packing efficiency of a β subunit, which is strongly dependent on the chemical compound bound to the β subunit, influences the packing efficiency of the subcomplex including the β subunit; (2) the nonuniformity of the packing structure of the $\alpha_3\beta_3$ complex is sufficiently high; (3) the hydration entropy of the $\alpha_3\beta_3\gamma$ complex is quite variable depending on the orientation of the γ subunit; and (4) a very large increase in the water-entropy loss for one of the three subcomplexes is almost cancelled out by large decreases in the water-entropy loss for the other two subcomplexes, causing no free-energy barrier for the rotation of the γ subunit.

None of the four requirements is completely met by the mutant. The absolute value of hydration entropy of state (c) or (d) in Fig. 3.23 is not much larger than that of state (a) or (b) shown in Fig. 3.22. In other words, for the mutant, the loss of water entropy caused in state change (a)→(c) or (a)→(d) is much smaller, and factor 2 is as significant as or more significant than factor 1 (see Sect. 3.8.1 for the definition of factors 1 and 2). Therefore, the ATP hydrolysis can persist even when the inverse rotation of the γ subunit is forcibly executed. Therefore, state change (a)→(c), where the ATP binding, ATP hydrolysis, and dissociation of ADP and Pi lead a decrease in system free energy, is often favored. For the wild type, during the rotation, a very large increase in the water-entropy loss for one of the three subcomplexes is almost cancelled out by large decreases in the water-entropy loss for the other two subcomplexes, giving rise to no free-energy barrier for the rotation of the γ subunit. This may not be the case for the mutant, resulting in a nontrivial free-energy barrier. For the mutant, overcoming this free-energy barrier by the thermal fluctuation is a stochastic process, causing the quite irregular rotational behavior where an abrupt decrease in the rotation rate is often encountered. It is not rare that the state change does not reach a stable state when the external torque is applied, and the inverse rotation can be accompanied by no reactions with the occurrence of state change (a)→(d) (see Figs. 3.22 and 3.23). Taken together, state changes (a)→(c) and (a)→(d) dominate for the mutant. The free-energy balance as that considered in Sect. 3.8.3 is no more valid. We can state for the mutant that the entropic force by water driving the normal rotation is considerably weaker and readily yields to the external force driving the inverse rotation.

It is remarkable that the thermodynamic behavior of F_1-ATPase is substantially influenced by the mutation E190D. However, portions closely packed like a three-dimensional jigsaw puzzle can be perturbed to a drastic extent even by a single mutation, giving rise to a significant modification of the packing structure of the whole protein complex. This modification may be responsible for the deterioration of the ATPase activity [36] and the lengthened catalytic dwell [24].

3.8.5 F_oF_1-ATP Synthase

F_oF_1-ATP synthase [37] synthesizes ATP in virtually all cells. F_o is within the membrane and F_1 is in the intracellular region. The γ subunit, which connects F_o and F_1, is incorporated in both of them. F_1-ATPase, the $\alpha_3\beta_3\gamma$ complex separated from F_oF_1-ATP synthase, hydrolyses ATP. In $F_o-\gamma$, sufficiently large W (W is an energy given to the system; see Eq. (3.2)) is realized by the coupling with the transfer of protons from the higher-proton-concentration (extracellular) region to the lower-proton-concentration (intracellular) region, an irreversible process accompanied by a free-energy decrease of $\sim -9k_BT$ ($T = 298$ K) [37]. As a result, the γ subunit rotates in the inverse direction in F_1 with the occurrence of ATP synthesis. A transfer of three protons is required for synthesizing one ATP molecule.

References

1. Uchihashi T, Iino R, Ando T, Noji H (2011) Science 333:755
2. Shirakihara Y, Yohda M, Kagawa Y, Yokoyama K, Yoshida M (1991) J Biochem 109:466
3. Bowler MW, Montgomery MG, Leslie AGW, Walker JE (2007) J Biol Chem 282:14238
4. Abrahams JP, Leslie AG, Lutter R, Walker JE (1994) Nature 370:621
5. Kabaleeswaran V, Shen H, Symersky J, Walker JE, Leslie AGW, Mueller DM (2009) J Biol Chem 284:10546
6. Okuno D, Fujisawa R, Iino R, Hirono-Hara Y, Imamura H, Noji H (2008) Proc Natl Acad Sci USA 105:20722
7. Sieladd H, Rennekamp H, Engelbrecht S, Junge W (2008) Biophys J 95:4979
8. Masaike T, Koyama-Horibe F, Oiwa K, Yoshida M, Nishizaka T (2008) Nat Struct Mol Biol 15:1326
9. Watanabe R, Iino R, Noji H (2010) Nat Chem Biol 6:814
10. Ito Y, Ikeguchi M (2010) J Comput Chem 31:2175
11. Šali A, Blundell TLJ (1993) J Mol Biol 234:779
12. Gibbons C, Montgomery MG, Leslie AGW, Walker JE (2000) Nat Struct Biol 7:1055
13. Yoshidome T, Ito Y, Ikeguchi M, Kinoshita M (2011) J Am Chem Soc 133:4030
14. Kusalik PG, Patey GN (1988) J. Chem. Phys. 88:7715
15. Kusalik PG, Patey GN (1988) Mol Phys 65:1105
16. Cann NM, Patey GN (1997) J Chem Phys 106:8165
17. Kinoshita M (2008) J Chem Phys 128:024507
18. Hayashi T, Oshima H, Harano Y, Kinoshita M (2016) J Phys: Condens Matter 28:344003
19. Roth R, Harano Y, Kinoshita M (2006) Phys Rev Lett 97:078101
20. Oshima H, Kinoshita M (2015) J Chem Phys 142:145103

21. Hayashi T, Inoue M, Yasuda S, Petretto E, Škrbić T, Giacometti A, Kinoshita M (2018) J Chem Phys 149:045105
22. Yasuda R, Noji H, Kinosita K Jr, Yoshida M (1998) Cell 93:1117
23. Yasuda R, Noji H, Yoshida M, Kinosita K Jr, Itoh H (2001) Nature 410:898
24. Shimabukuro K, Yasuda R, Muneyuki E, Hara KY, Kinosita K Jr, Yoshida M (2003) Proc Natl Acad Sci USA 100:14731
25. Adachi K, Oiwa K, Nishizaka T, Furuike S, Noji H, Itoh H, Yoshida M, Kinosita K Jr (2007) Cell 130:309
26. Furuike S, Hossain MD, Maki Y, Adachi K, Suzuki T, Kohori A, Itoh H, Yoshida M, Kinosita K Jr (2008) Science 319:955
27. Ma J, Flynn TC, Cui Q, Leslie AGW, Walker JE, Karplus M (2002) Structure 10:921
28. Hara KY, Noji H, Bald D, Yasuda R, Kinosita K Jr, Yoshida M (2000) J Biol Chem 275:14260
29. Tanigawara M, Tabata KV, Ito Y, Ito J, Watanabe R, Ueno H, Ikeguchi M, Noji H (2012) Biophys J 103:970
30. Kabaleeswaran V, Puri N, Walker JE, Leslie AGW, Mueller DM (2006) EMBO J 25:5433
31. Yoshidome T, Ito Y, Matubayasi N, Ikeguchi M, Kinoshita M (2012) J Chem Phys 137:035102
32. Muneyuki E, Watanabe-Nakayama T, Suzuki T, Yoshida M, Nishizaka T, Noji H (2007) Biophys J 92:1806
33. Toyabe S, Watanabe-Nakayama T, Okamoto T, Kudo S, Muneyuki E (2011) Proc Natl Acad Sci USA 108:17951
34. Toyabe S, Muneyuki E (2015) New J Phys 17:015008
35. Tanaka M, Kawakami T, Okaniwa T, Nakayama Y, Toyabe S, Ueno H, Muneyuki E (2020) Biophys J 119:48
36. Hayashi S, Ueno H, Shaikh AR, Umemura M, Kamiya M, Ito Y, Ikeguchi M, Komoriya Y, Iino R, Noji H (2012) J Am Chem Soc 134:8447
37. Voet D, Voet JG (2004) Biochemistry, 3rd edn. Wiley, New York

Chapter 4
Concluding Remarks

Abstract AcrB, a homotrimer possessing a triangular-prism shape, is the principal part of the AcrA–AcrB–TolC complex which extrudes a variety of drugs from a cell. AcrB comprises three protomers which are in access (A), binding (B), and extrusion (E) states along a drug-transport cycle, respectively. According to a suggestion made in the literature, the three protomers exhibit a sequential conformational change expressed as (A, B, E)→(B, E, A)→(E, A, B)→(A, B, E). This change, which is referred to as the "functional rotation", is achieved with the use of the so-called proton motive force yielding repeated proton binding to and dissociation from AcrB. In this chapter, we point out that F_1-ATPase considered in Chap. 3 and AcrB share physically the same rotation mechanism. Whenever the structure of one of the three portions forming a protein complex is perturbed in the direction that a solvent-entropy loss is caused, the structures of the other two portions are reorganized to make up for the loss. We also comment on the rotation mechanism of V_1-ATPase which we intend to explore in the next stage.

Keywords AcrB · Proton motive force · Drug extrusion · Multidrug efflux · Functional rotation · V_1-ATPase

4.1 Functional Rotation of AcrB

In Chap. 3, we have discussed the functional expression of F_1-ATPase, the unidirectional rotation of the γ subunit. The important points can be summarized as follows: The system of interest comprises not only F_1-ATPase but also water in which ATP, ADP, and Pi are dissolved (water is not the external system for F_1-ATPase); F_1-ATPase is coupled with the ATP hydrolysis reaction, an irreversible process accompanied by a decrease in system free energy; F_1-ATPase is thus involved in the ATP hydrolysis cycle (i.e., the ATP binding to F_1-ATPase, ATP hydrolysis, and dissociation of ADP and Pi from F_1-ATPase) which spontaneously occurs; the system performs essentially no mechanical work during the rotation; and the force rotating the γ subunit is generated by not ATP but water. The force originates from the translational displacement of water molecules in the whole system.

© The Author(s), under exclusive license to Springer Nature Singapore Pte Ltd. 2021   63
M. Kinoshita, *Mechanism of Functional Expression of F_1-ATPase*,
SpringerBriefs in Molecular Science,
https://doi.org/10.1007/978-981-33-6232-1_4

There is another class of proteins or protein complexes utilizing the proton motive force characterized by the repeated transfer of a proton from the higher-concentration side to the lower-concentration one. A paradigmatic example is the drug efflux pump, the *Escherichia coli* AcrA–AcrB–TolC tripartite complex, which is formed by the polytopic inner membrane protein AcrB, periplasmic adaptor protein AcrA, and outer membrane channel TolC [1–4]. The complex extrudes a variety of drugs from a cell. AcrB is a homotrimer possessing a triangular-prism shape and in charge of the major part of pumping drugs out of the cell from the inner membrane or periplasm via the TolC channel (see Fig. 4.1). As shown in Fig. 4.2, each protomer of AcrB consists of the transmembrane (TM), porter, and drug-efflux domains and takes three distinct

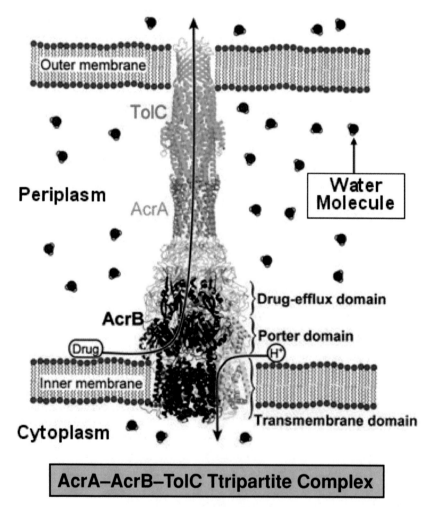

Fig. 4.1 Schematic diagram of AcrA–AcrB–TolC tripartite complex (Adapted with permission from Ref. 11. Copyright 2015 American Chemical Society)

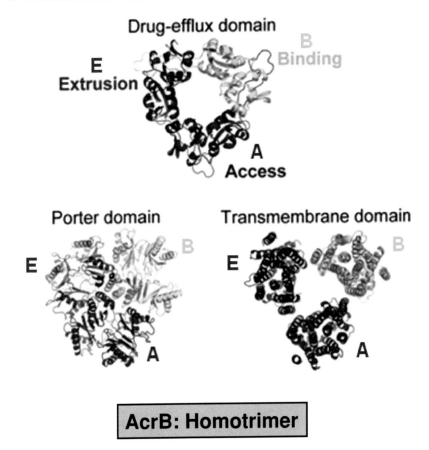

Fig. 4.2 Schematic diagram of AcrB comprising three protomers (see Fig. 4.1). This figure is viewed from the top. The protomers in the access (A), binding (B), and extrusion (E) states are drawn in blue, yellow, and red, respectively. Each protomer is constructed by the transmembrane (TM), porter, and drug-efflux domains. (Adapted with permission from Ref. 11. Copyright 2015 American Chemical Society)

structures which are in access (A), binding (B), and extrusion (E) states along a drug-transport cycle, respectively. In the literature, we find a "functionally rotating" picture in which each protomer exhibits a sequential conformational change expressed as (A, B, E)→(B, E, A)→(E, A, B)→(A, B, E) [2]. This change is achieved through repeated proton binding to and dissociation from AcrB. It was suggested that a proton binds to Asp408 in the TM domain of one of the three protomers and the proton translocation stoichiometry is a single proton per cycle, (A, B, E)→(B, E, A) [5]. The transfer of a proton from the higher-concentration side to the lower-concentration one is an irreversible process accompanied by a decrease in system free energy. The functional rotation, which is involved in this process, spontaneously occurs.

The porter and drug-efflux domains are immersed in water molecules but the TM domain is immersed in nonpolar chains of lipid molecules. These nonpolar chains as well as water molecules act as the "solvent". We showed that the translational, configuration entropy of hydrocarbon groups in the nonpolar chains (CH_2, CH_3, and CH) and water molecules provide a clue to the structural stability of a membrane protein [6–10] and to the mechanism of the functional rotation of AcrB [11, 12]. We showed that the packing structure of AcrB with a proton or two protons bound is characterized by significant ununiformity and this ununiformity plays essential roles in the functional rotation through the solvent-entropy effect [11, 12]. There are similarities between the rotations for the $\alpha_3\beta_3$ complex in F_1-ATPase and AcrB. The solvent is not the external system for AcrB. The force required for the functional rotation of AcrB is generated by the solvent.

As argued in Sect. 3.3.2 through 3.3.4, in the most stable packing structure of the $\alpha_3\beta_3$ complex in F_1-ATPase, the packing efficiencies of subcomplexes I−γ, II−γ, and III−γ are substantially different from one another. Due to the ATP hydrolysis cycle, the most stable packing structure is perturbed and the structural reorganization of the $\alpha_3\beta_3$ complex occurs to retain the most stable packing structure. The physical essence can be described as follows: When the structure of one of the three portions forming the complex is perturbed in the direction that a solvent-entropy loss is caused, the structures of the other two portions are always reorganized to make up for the loss and thus prevent a decrease in solvent entropy. This physical essence should also be applicable to the functional rotation of AcrB. In AcrB, the three portions correspond to the protomers in access (A), binding (B), and extrusion (E) states, respectively. The solvent is composed of water molecules for F_1-ATPase while it is composed of the hydrocarbon groups mentioned above as well as water molecules for AcrB.

The functional rotation of AcrB suggested by us is depicted in Fig. 4.3. Consult our earlier publications [11, 12] for more details. Significant points are briefly described in what follows. The protomers in access (A), binding (B), and extrusion (E) states are referred to as protomers A, B, and E, respectively. A drug molecule is accommodated in protomer B, it is inserted during the structural reorganization from protomer A to protomer A', and it is extruded during the structural reorganization from protomer B' to protomer E. In conformation (a), only the proton binding site of protomer B is exposed to the higher-concentration side, and a proton binds to protomer B ((a)→(b)). The binding is accompanied by a decrease in system free energy. During the reorganization of the AcrB structure following the proton binding ((b)→(c)), a drug molecule is entropically introduced. The introduction leads to a solvent-entropy gain. The proton binding site of protomer E' is then exposed to the lower-concentration side with the result of the dissociation of the proton from protomer E' ((c)→(d)). The dissociation is accompanied by a decrease in system free energy. During the reorganization of the AcrB structure following the proton dissociation ((d)→(e)), the drug molecule is entropically released. The release leads to a solvent-entropy gain.

The net decrease in system free energy in each drug-transport cycle (or equivalently, each proton-transfer cycle) is F_P ($F_P < 0$). F_P is $\sim -9k_BT + \xi$ ($\xi > 0$).

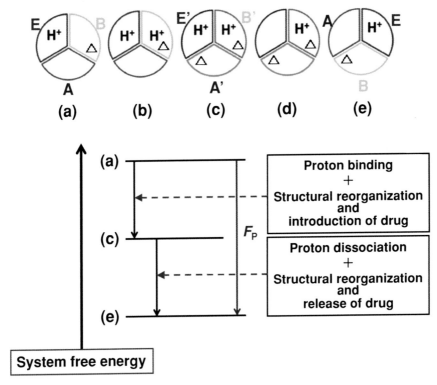

Fig. 4.3 Top: Conformational change of AcrB during one drug-transport cycle, or equivalently, one proton-transfer cycle. A different color represents a different state (access (A), binding (B), or extrusion (E) state). The open triangle denotes a drug molecule. (a) Start of proton-transfer cycle. (b) Right after a proton binds to protomer B. (c) After reorganization of AcrB structure accompanied by introduction of a drug molecule into protomer A (protomer A changes to protomer A' upon the reorganization). (d) Right after a proton dissociates from protomer E'. (e) After reorganization of AcrB structure accompanied by release of a drug molecule from protomer B' (protomer B' changes to protomer E upon the reorganization). End of proton-transfer cycle. The structures of AcrB in (a) and (e) are the same. The structural changes of (a)→(b)→(c)→(d)→(e)=(a) correspond to (A, B, E)→(A', B', E')→(B, E, A). Bottom: Decrease of system free energy during one drug-transport cycle. The net decrease is F_P (<0)

The transfer of a single proton from the higher-concentration side to the lower-concentration one leads to an increase in system entropy, and the transfer of a positive charge along the electrostatic-potential gradient results in a decrease in system energy, which sums to a free-energy decrease of $-9k_B T$. ξ is attributed to the pumping of a drug molecule from the lower-concentration side to the higher-concentration one [16]. In our view, F_1-ATPase and AcrB share physically the same rotation mechanism.

As discussed in Sect. 2.8, the entropic insertion of a solute into a vessel consisting of biopolymers followed by the entropic release of the solute from the vessel is a fundamental process in a biological system. Interestingly, after a solute is inserted into the vessel possessing a specific structure, the solute is released from the vessel upon a structural change of the vessel [12–15]. AcrA/AcrB/TolC is capable of handling drugs with diverse properties and therefore characterized by the "multidrug efflux". This characteristic is conferred upon the vessel only when the solvent-entropy effect dominates [11, 12, 15].

4.2   Toward Investigation of Unidirectional Rotation of Central Shaft in V_1-ATPase

Water-soluble V_1-ATPase, the catalytic domain of Vacuolar ATPase, is an ATP-driven molecular motor which appears to be quite similar to F_1-ATPase [17–22]. V_1-ATPase utilizes the ATP hydrolysis cycle. It comprises A, B, D, and F subunits, and three AB pairs are hexagonally arranged around the DF subcomplex. The A and B subunits and the DF subcomplex in V_1-ATPase correspond to the β, α, and γ subunits in F_1-ATPase, respectively. Under the solution condition that the ATP hydrolysis reaction occurs, V_1-ATPase hydrolyzes ATP into ADP and Pi, which is accompanied by the rotation of the DF subcomplex in the counterclockwise direction. V_1-ATPase has been investigated mostly by Murata and coworkers [17–22], and several differences between V_1-ATPase and F_1-ATPase in the structural characteristics and in the rotational mode have been pointed out. The mechanism of the functional expression of V_1-ATPase remains quite elusive. We intend to collaborate with the group of Murata for clarifying similarities and differences between F_1-ATPase and V_1-ATPase in the packing structure, chemical compounds bound to the three β or A subunits, and rotational mechanism. Despite the possible differences, it is definite that the water-entropy effect plays a pivotal role for V_1-ATPase as well and the rotational mechanisms of F_1-ATPase and V_1-ATPase share the same physical essence. For V_1-ATPase, the crystal structures in significantly more different states have been solved by experiments. The collaboration mentioned above is expected to not only unveil the rotational mechanism of V_1-ATPase but also make some of the arguments for F_1-ATPase described in Chap. 3 even more convincing or, if necessary, modify the details of its rotational mechanism.

References

1. Koronakis V, Sharff A, Koronakis E, Luisi B, Hughes C (2000) Nature 405:914
2. Murakami S, Nakashima R, Yamashita E, Matsumoto T, Yamaguchi A (2006) Nature 443:173
3. Seeger MA, Schiefner A, Eicher T, Verrey F, Diederichs K, Pos KM (2006) Science 313:1295
4. Sennhauser G, Amstutz P, Briand C, Storchenegger O, Grütter M (2007) PLOS Biol. 5:e7(0106)
5. Yamane T, Murakami S, Ikeguchi M (2013) Biochemistry 52:7648
6. Yasuda S, Kajiwara Y, Takamuku Y, Suzuki N, Murata T, Kinoshita M (2016) J Phys Chem B 120:3833
7. Yasuda S, Kajiwara Y, Toyoda Y, Morimoto K, Suno R, Iwata S, Kobayashi T, Murata T, Kinoshita M (2017) J Phys Chem B 121:6341

8. Yasuda S, Hayashi T, Kajiwara Y, Murata T, Kinoshita M (2019) J Chem Phys 150:055101
9. Yasuda S, Kazama K, Akiyama T, Kinoshita M, Murata T (2020) J Mol Liq 301:112403
10. Murata T, Yasuda S, Hayashi T, Kinoshita M (2020) Biophys Rev 12:323
11. Mishima H, Oshima H, Yasuda S, Kinoshita M (2015) J Phys Chem B 119:3423
12. Kinoshita M (2016) Mechanism of functional expression of the molecular machines. Springer Briefs in Molecular Science, Springer, ISBN: 978-981-10-1484-0
13. Amono K, Kinoshita M (2010) Chem Phys Lett 488:1
14. Mishima H, Oshima H, Yasuda S, Amano K, Kinoshita M (2013) Chem Phys Lett 561–562:159
15. Mishima H, Oshima H, Yasuda S, Amano K, Kinoshita M (2013) J Chem Phys 139:205102
16. Voet D, Voet JG (2004) Biochemistry, 3rd edn. Wiley, New York
17. Murata T, Yamato I, Kakinuma Y, Leslie AGW, Walker JE (2005) Science 308:654
18. Murata T, Yamato I, Kakinuma Y, Shirouzu M, Walker JE, Yokoyama S, Iwata S (2008) Proc Natl Acad Sci USA 105:8607
19. Mizutani K, Yamamoto M, Suzuki K, Yamato I, Kakinuma Y, Shirouzu M, Walker JE, Yokoyama S, Iwata S, Murata T (2011) Proc Natl Acad Sci USA 108:13474
20. Saijo S, Arai S, Hossain KMM, Suzuki K, Yamato I, Kakinuma Y, Ishizuka-Katsura Y, Ohsawa N, Terada T, Shirouzu M, Yokoyama S, Iwata S, Murata T (2011) Proc Natl Acad Sci USA 108:19955
21. Arai S, Saijo S, Suzuki K, Mizutani K, Kakinuma Y, Ishizuka-Katsura Y, Ohsawa N, Terada T, Shirouzu M, Yokoyama S, Iwata S, Yamato I, Murata T (2013) Nature 493:703
22. Iida T, Minagawa Y, Ueno H, Kawai F, Murata T, Iino R (2019) J Biol Chem 294:17017

Chapter 5
Appendix 1: Angle-Dependent Integral Equation Theory

Abstract By virtue of the recently achieved, remarkable progress of the integral equation theories for solute hydration, we can now handle a large biomolecule like a protein or protein complex immersed in water by means of statistical mechanics in which the biomolecule structure is taken into account at the atomic level and a molecular model is employed for water. The angle-dependent integral equation (ADIE) theory is best suited to the assessment of the hydrophobic hydration, the most important physical factor in hydration thermodynamics. In this chapter, we briefly summarize the basic characteristics of the ADIE theory.

Keywords Integral equation theory · Ornstein-Zernike relation · Closure equation · Angle-dependent integral equation theory · Hydration thermodynamics · Hydrophobic hydration

The Ornstein-Zernike (OZ) relation and a closure equation are the two basic equations in an integral equation theory (IET) [1]. They are derived from the system partition function using a variety of correlation and distribution functions defined on the basis of classical statistical mechanics. Not only the solvent structure near a solute but also thermodynamic quantities of solvation (e.g., the solvation free energy, energy, and entropy) can be calculated via the following two steps:

Step 1. The bulk solvent is treated. In the case of a single component, the temperature, number density, and solvent-solvent interaction potential form the input data. By numerically solving the two basic equations mentioned above, we can calculate the solvent-solvent correlation functions, microscopic density and orientational structures of the solvent, and thermodynamic quantities. Unlike in the molecular dynamics (MD) and Monte Carlo (MC) simulations, the average value of a physical quantity can be calculated, in essence, for an infinitely large system and an infinitely large number of system configurations. The IET is free from the problems of too small a system size and statistical errors, which often arise in the MD and MC simulations. Not only a simple-fluid solvent but also a molecular liquid such as water can be treated by the IET. For water, the angle-dependent integral equation (ADIE) theory [2–6] is the most reliable tool.

© The Author(s), under exclusive license to Springer Nature Singapore Pte Ltd. 2021   71
M. Kinoshita, *Mechanism of Functional Expression of F₁-ATPase*,
SpringerBriefs in Molecular Science,
https://doi.org/10.1007/978-981-33-6232-1_5

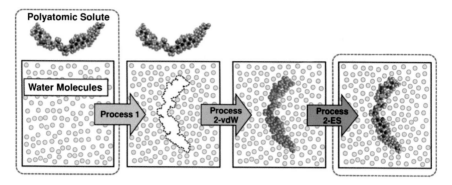

Fig. 5.1 Decomposition of hydration of polyatomic solute. It can be decomposed into processes 1 and 2 [7–9]. Process 1 is the hydrophobic hydration. Process 2 can further be decomposed into processes 2-vdW and 2-ES

Step 2. The system treated is a solute inserted into the solvent in step 1 under the isochoric condition at the infinite dilution limit. The solute-solvent correlation functions are calculated using the solvent-solvent correlation functions and the solute-solvent interaction potential as the input data. For a spherical solute, the ADIE theory [2–6] is best suited to an analysis of solute solvation when the solvent is water (i.e., an analysis of solute hydration).

The hydration of a solute molecule (e.g., a protein) with a fixed structure can be decomposed into the following two processes [7–9] (see Fig. 5.1).

Process 1. Creation of a cavity in water. The cavity matches the polyatomic structure of the solute molecule. The cavity is modeled as a set of fused, neutral hard spheres corresponding to the atoms constituting the solute molecule. The diameter of each neutral hard sphere is set at one of the Lennar-Jones (LJ) potential parameters, σ, assigned to each atom. Process 1 is the hydrophobic hydration.

Process 2. Incorporation of solute-water van der Waals (vdW) potential followed by that of solute-water electrostatic (ES) potential. Process 2 is composed of processes 2-vdW and 2-ES. In process 2-vdW, the hard-sphere repulsive potential between an atom in the solute molecule and a water molecule (i.e., the atom-water hard-sphere repulsive potential) is replaced by the LJ potential. In process 2-ES, a prescribed partial charge is given to each atom in the solute molecule to incorporate the atom-water ES potential.

It was shown in our earlier works [8–10] that the hydration properties relevant to process 1 are much more significant than those relevant to process 2 in arguing such processes as the protein-peptide binding [8] and the protein denaturation [9, 10]. The hydration properties relevant to process 1, which depends on the temperature and the pressure much more strongly than those relevant to process 2, plays essential roles in the cold [9] and pressure [10] denaturating of a protein and in the globule-to-coil transition poly(N-isopropylacrylamide) (PNIPAM) [9] caused at low temperatures. In particular the hydration entropy in process 1 is the most important quantity. We note that the hydration entropy in process 2 is much smaller than that in process 1.

The hydration entropy of a polyatomic solute in process 1 can be calculated with sufficient accuracy and very high speed by the ADIE theory combined with the morphometric approach [11–13] (see Chap. 6).

In the ADIE theory [2–6], the dependences of interaction potential and a correlation function on the orientations of water molecules are explicitly taken into account. As a molecular model for water, a multipolar model [3] is the most conveniently employed. Here, we assume that a neutral hard sphere with diameter d_U is considered as the solute (see step 2 described above). The OZ equation is expressed as

$$h(12) = c(12) + \{1/(8\pi^2)\}\rho_s \int c(13)h_{ss}(32)d(3), \qquad (5.1)$$

where h is the solute-water total correlation function, c is the solute-water direct correlation function, ρ_S is the number density of bulk water, h_{SS} is the water-water total correlation function, (ij) signifies $(\boldsymbol{\Omega}_i, \boldsymbol{\Omega}_j, \boldsymbol{r}_{ij})$ where $\boldsymbol{\Omega}_i$ represents the three Euler angles describing the orientation of particle i and $\boldsymbol{r}_{ij}$ is the vector connecting the centers of particles i and j, and $d(3)$ denotes the integration over all position and angular coordinates of particle 3. When the solute (particle 1) is a neutral hard sphere, $h(12)$ or $c(12)$ is dependent on $\boldsymbol{\Omega}_2$ and $\boldsymbol{r}_{12}$. It should be emphasized that the OZ equation is formally exact. The closure equation is expressed as

$$C(12) = \int_r^\infty [h(12)\partial\{w(12) - b(12)\}/\partial r']dr' - u(12)/(k_BT) + b(12), \quad (5.2)$$

$$w(12) = c(12) - h(12) + u(12)\big/(k_BT), \qquad (5.3)$$

where b is the bridge function, u is the interaction potential, and r is the distance between the centers of two particles. The closure equation is reformulated so that the rotational-invariant expansion described below can be applied to it. The application of the self-consistent mean field (SCMF) theory [2, 3] enables us to take account of the effect of molecular polarizability of water. In this theory, the many-body induced interactions are reduced to pairwise additive potentials involving an effective dipole moment. This effective dipole moment determined is about 1.42 times larger than the bare gas-phase dipole moment at 298 K and 1 atm. We showed that the HNC approximation ($b = 0$) gives quite accurate results [6].

Since the two basic equations (i.e., the OZ relation and the HNC closure) include up to 6-variable functions and 6-fold integrations, they are not numerically tractable in their original forms. For the pragmatic numerical solution of the two basic equations, a correlation function is expanded in a basis set of rotational invariants. The two basic equations are then reformulated using the projections $X^{mnl}_{\mu\nu}$ in the rotational-invariant expansion of a water-water or water-solute correlation function X [2–6]. Our experience showed that the expansion considered for $m, n \le n_{max} = 4$ gives sufficiently accurate results for a nonpolar solute immersed in water. r_L, which is chosen such that the correlations at $r = r_L$ become sufficiently weak, is discretized into N grid points ($r_i = i\delta r$, i=0, 1, ..., $N-1$; $\delta r = r_L/N$; $\delta r = 0.01d_S$; $N = 4096$), and all the projections are represented by their values on these points. The numerical solution is performed using the robust, highly efficient algorithm developed by Kinoshita and coworkers [14, 15]. The hydration entropy of the solute, S, is evaluated via the temperature derivative of hydration free energy μ calculated using the Morita-Hiroike formula extended to a molecular liquid (V is the system volume) []:

$$S = -(\partial\mu/\partial T)_V \sim -\{\mu(T + \Delta T) - \mu(T + \Delta T)\}/(2\Delta T), \Delta T = 5K. \quad (5.4)$$

In general, the dielectric constant of water is a good measure of the validity of a theory. The ADIE theory gives a value of ~83 that is in very close agreement with the experimental value (~78).

In the three-dimensional reference interaction site model (3D-RISM) theory [16–19], another IET for hydration of a polyatomic solute, the mathematical complications are avoided using an approximation in which the dependence of a correlation function on the orientations of water molecules is not explicitly taken into account. Due to this approximation, the OZ equation as well as the closure equation is not exact [6]. The most serious drawback of the 3D-RISM theory is that it completely fails to elucidate the hydrophobic hydration (i.e., process 1) [6]. For a nonpolar solute, even when a realistic molecular model is employed for water, the water behaves like a hard-sphere solvent (the number density and the particle diameter are equal to those pertinent to water). As a consequence, the 3D-RISM theory predicts that the hydrophobicity is strengthened at low temperatures. This prediction is opposite to the experimental evidence that the hydrophobicity is weakened at low temperatures [5, 6, 9]. Therefore, it is incapable of elucidating such subjects as the cold denaturation of a protein. It also gives too high a hydration free energy of a nonpolar solute and too high a density profile of water near a nonpolar solute [6]. However, the 3D-RISM theory can suitably be applied to an analysis on process 2 (see Chap. 6).

References

1. Hansen J-P, McDonald LR (2006) Theory of simple liquids, 3rd edn. Academic, London
2. Kusalik PG, Patey GN (1988) J Chem Phys 88:7715
3. Kusalik PG, Patey GN (1988) Mol Phys 65:1105
4. Cann NM, Patey GN (1997) J Chem Phys 106:8165
5. Kinoshita M (2008) J Chem Phys 128:024507
6. Hayashi T, Oshima H, Harano Y, Kinoshita M (2016) J Phys: Condens Matter 28:344003
7. Hikiri S, Hayashi T, Inoue M, Ekimoto T, Ikeguchi M, Kinoshita M (2019) J. Chem. Phys. 150:175101
8. Yamada T, Hayashi T, Hikiri S, Kobayashi N, Yanagawa H, Ikeguchi M, Katahira M, Nagata T, Kinoshita M (2019) J Chem Inf Model 59:3533
9. Inoue M, Hayashi T, Hikiri S, Ikeguchi M, Kinoshita M (2020) J Mol Liq 317:114129
10. Inoue M, Hayashi T, Hikiri S, Ikeguchi M, Kinoshita M (2020) J Chem Phys 152:065103
11. Roth R, Harano Y, Kinoshita M (2006) Phys Rev Lett 97:078101
12. Oshima H, Kinoshita M (2015) J Chem Phys 142:145103

13. Hayashi T, Inoue M, Yasuda S, Petretto E, Škrbić T, Giacometti A, Kinoshita M (2018) J Chem Phys 149:045105
14. Kinoshita M, Harada M (1991) Mol Phys 74:443
15. Kinoshita M, Bérard DR (1996) J Comp Phys 124:230
16. Ikeguchi M, Doi J (1995) J Chem Phys 103:5011
17. Beglov D, Roux B (1995) J Chem Phys 103:360
18. Beglov D, Roux B (1996) J Chem Phys 104:8678
19. Kovalenko A, Hirata F (1999) J Chem Phys 110:10095

Chapter 6
Appendix 2: Morphometric Approach

Abstract The hydration of a biomolecule with a fixed structure can be decomposed into the following two processes: process 1, the hydrophobic hydration, where a cavity matching the polyatomic structure of the biomolecule is created; and process 2 where biomolecule-water van der Waals and electrostatic interaction potentials are taken into account. In process 1, the cavity is treated as a solute in the angle-dependent integral equation (ADIE) theory, the most reliable statistical-mechanical theory for solute hydration. However, such a complexly shaped solute cannot directly be handled by the ADIE theory because of the mathematical complications encountered. We have solved this problem by combining the ADIE theory with the morphometric approach (MA). In this chapter, we briefly summarize the basic characteristics of the MA. Our new hybrid method where this ADIE-MA theory and the three-dimensional reference interaction site model (3D-RISM) theory are applied to processes 1 and 2, respectively, is capable of calculating the hydration free energy, energy, and entropy of a large polyatomic solute like a protein with sufficient accuracy and high speed.

Keywords Cavity creation · Hydrophobic effect · van der Waals potential · Electrostatic potential · Morphometric approach · Excluded volume · Water-accessible surface area

The cavity created in process 1, which matches the polyatomic structure of a solute molecule, is modeled as a set of fused, neutral hard spheres. In the morphometric approach (MA) [1–3], the cavity is geometrically characterized by the excluded volume V_{ex}, water-accessible surface area A, and integrated mean and Gaussian curvatures of the accessible surface denoted by Y and Z, respectively. A thermodynamic quantity of hydration in process 1, X_{p1}, is expressed as the linear combination of the four geometric measures (i.e., V_{ex}, A, Y, and Z):

$$X_{p1} = C_1(X_{p1})V_{ex} + C_2(X_{p1})A + C_3(X_{p1})Y + C_4(X_{p1})Z. \qquad (6.1)$$

© The Author(s), under exclusive license to Springer Nature Singapore Pte Ltd. 2021
M. Kinoshita, *Mechanism of Functional Expression of F$_1$-ATPase*,
SpringerBriefs in Molecular Science,
https://doi.org/10.1007/978-981-33-6232-1_6

Here, X_{p1} is $\mu_{H,1}/(k_BT)$, $\varepsilon_{VH,1}/(k_BT)$, or $S_{VH,1}/k_B$ ($\mu_{H,1}$, $\varepsilon_{VH,1}$, and $S_{VH,1}$ are the hydration free energy, energy, and entropy, respectively, k_B is the Boltzmann constant, and T is the absolute temperature). Equation (6.1) is referred to as the "morphometric form". The four coefficients, $C_1(X_{p1})$–$C_4(X_{p1})$, are dependent only on the thermodynamic state of bulk water. Hence, they are determined for isolated, spherical cavities possessing much simpler geometric properties as described below.

Beforehand, we calculate values of X_{p1} for isolated neutral hard spheres with sufficiently many different diameters using the ADIE theory [4–8]. By applying the morphometric form for the isolated neutral hard spheres to the sufficiently many combinations of the values of X_{p1} calculated and the diameter, we determine $C_1(X_{p1})$–$C_4(X_{p1})$ by means of the least-squares method. Once $C_1(X_{p1})$–$C_4(X_{p1})$ are determined, all one has to do is to calculate V_{ex}, A, Y, and Z of the cavity from the Cartesian coordinates of the center of each neutral hard sphere in the cavity and its diameter σ (σ is a Lennard-Jones potential parameter assigned to each neutral hard sphere which corresponds to each atom in the polyatomic solute). X_{p1} of the cavity is then obtained from Eq. (6.1). The MA is advantageous in the following two respects: (i) X_{p1} can be calculated with very high speed (in ~1 s even for a large protein with a prescribed structure on a standard workstation); and (ii) X_{p1} can be decomposed into a variety of physically insightful components, and by assessing their signs and relative magnitudes we can disclose the dependences of hydration properties on the geometric characteristics of a polyatomic solute, temperature, and pressure. Consult our earlier publications [2, 9–13] for more details.

We have recently developed a new method [9, 13] for calculating the hydration free energy, energy, and entropy of a solute molecule. The solute hydration can be decomposed into processes 1 and 2 explained in Chap. 5. In the new method, the ADIE theory combined with the MA is employed for process 1, and the 3D-RISM theory is applied to process 2. The new method enables us to finish the calculation for a large polyatomic solute like a protein with sufficient accuracy and high speed. (Most of the computation time is consumed for process 2 where the 3D-RISM theory is employed.) Solutes with a wide range of sizes can be handled in the same manner. Neither a stage of training nor parameterization is necessitated. A solute possessing a significantly large total charge can be handled without difficulty.

We note that the free-energy perturbation [14–16] and thermodynamic integration [17] methods using the molecular dynamics (MD) simulations suffer an unacceptably heavy computational burden [18, 19]. This problem was solved by the development of a novel MD simulation method based on solution theory in energy representation (the so-called ER method) [20–22]. However, the ER method is inapplicable to a solute possessing a significantly large total charge. We also note that the decomposition of the hydration free energy, energy, and entropy into a variety of physically insightful components is impractical in the MD simulations. It is worthwhile to add that values of the hydration free energy calculated by our new method for many different proteins with zero total charge were compared to those obtained using the ER method, and the agreement was just excellent.

We conclude this book by noting the following. In our new method, a thermodynamic quantity of hydration is calculated for a fixed structure of the solute. However, this causes no problems at all. The structural fluctuation of the solute in water can readily be taken into account in accordance with the following two steps [13]: (I) An ensemble consisting of sufficiently many solute structures is generated by a short MD simulation in explicit aqueous solution; and (II) a thermodynamic quantity of hydration is calculated for all the structures generated, and it is determined as the average value. In the MD simulation, the generation of the ensemble is much less time consuming than the calculation of the thermodynamic quantity of hydration. It was shown, for instance, that the relative values of the binding free energy calculated for an oncoprotein MDM2 and two different peptides (the reference peptide is the extreme N-terminal peptide region of a tumor suppressor protein p53 (p53NTD)) by employing the two steps explained above are in quantitatively good agreement with those obtained by experiments [10].

References

1. Roth R, Harano Y, Kinoshita M (2006) Phys Rev Lett 97:078101
2. Oshima H, Kinoshita M (2015) J Chem Phys 142:145103
3. Hayashi T, Inoue M, Yasuda S, Petretto E, Škrbić T, Giacometti A, Kinoshita M (2018) J Chem Phys 149:045105
4. Kusalik PG, Patey GN (1988) J Chem Phys 88:7715
5. Kusalik PG, Patey GN (1988) Mol Phys 65:1105
6. Cann NM, Patey GN (1997) J Chem Phys 106:8165
7. Kinoshita M (2008) J Chem Phys 128:024507
8. Hayashi T, Oshima H, Harano Y, Kinoshita M (2016) J Phys: Condens Matter 28:344003
9. Hikiri S, Hayashi T, Inoue M, Ekimoto T, Ikeguchi M, Kinoshita M (2019) J Chem Phys 150:175101
10. Yamada T, Hayashi T, Hikiri S, Kobayashi N, Yanagawa H, Ikeguchi M, Katahira M, Nagata T, Kinoshita M (2019) J Chem Inf Model 59:3533
11. Inoue M, Hayashi T, Hikiri S, Ikeguchi M, Kinoshita M (2020) J Chem Phys 152:065103
12. Inoue M, Hayashi T, Hikiri S, Ikeguchi M, Kinoshita M (2020) J Mol Liq 317:114129
13. Kinoshita M, Hayashi T (2020) Biophys Rev 12:469
14. Postma JPM, Berendsen HJC, Haak JR (1982) Faraday Symp Chem Soc 17:55
15. Tembe BL, McCammon JA (1984) Comput Chem 8:281
16. Jorgensen WL, Ravimohan C (1985) J Chem Phys 83:3050
17. Kirkwood JG (1935) J. Chem. Phys. 3:300
18. Kollman P (1993) Chem Rev 93:2395
19. Shirts MR, Pitera JW, Swope WC, Pande VS (2003) J Chem Phys 119:5740
20. Matubayasi N, Nakahara M (2000) J Chem Phys 113:6070
21. Matubayasi N, Nakahara M (2003) J Chem Phys 117:3605 (2002); 118:2446 (2003)
22. Matubayasi N, Nakahara M (2003) J Chem Phys 119:9686

Printed in the United States
by Baker & Taylor Publisher Services